Patrick Moore's Practical Astronomy Series

W0035992

For further volumes:
http://www.springer.com/series/3192

Classic Telescopes

A Guide to Collecting, Restoring, and Using Telescopes of Yesteryear

Neil English

 Springer

Neil English
25 Culcreuch Avenue
Fintry, Glasgow
G63 0YB, UK

ISSN 1431-9756
ISBN 978-1-4614-4423-7   ISBN 978-1-4614-4424-4 (eBook)
DOI 10.1007/978-1-4614-4424-4
Springer New York Heidelberg Dordrecht London

Library of Congress Control Number: 2012941614

Printed on acid-free paper

Springer is part of Springer Science+Business Media (www.springer.com)

The War is over.
I am at peace.
A New Empire of thought is established, where the lowly are lifted
and the humble can take heart.
They shall cower no more, new champions of the heavens emerging
on every shore.
'tis a kingdom of memes, auguring an Antonine Age, that illuminates
the life of the learned astronomer of yore,
like never before,
whose lonely vigil on hill and mountain high, through freezing
winter nights and sweltering summer days, was taken to the grave.

Raise a spyglass to Huygens and Huggins, to Bessel, Burnham and
Barnard.
To the Mozart of practical optics, whose death was premature,
to British Cooke and American Clark - the Special Relationship was
there for sure.
Three Cheers for Piazzi & Peltier, surveyors of the sky,
and for myopic Dawes, whose noble truths were first received
with wry.
The shadows cast by these great men will never fade from view.
They led the way, through brilliant dark, to fertile pastures new.
And when I take my spyglass and turn it to the sky,
I know for sure they saw enough, everything!, to soothe a weary Eye.
The War is over.
I am at peace.

Words dedicated to the memory of my late father, John J. English
(1923–2012)

Acknowledgments

Any book as ambitious as this one could never have seen the light of day without the generous contributions of so many people, who have so graciously devoted some of their spare time to share their experiences of their classic 'scopes. A big thank you to John Nanson, Bob Abraham, Jeff Morgan, Pat Conlon, Dick Parker, Jim Barnett, Richard Day, Gerald Morris, Steve Collingwood, Douglas Daniels, Bob Argyle, Phil Jaworek, Erik Bakker, Bill Nielsen, Alexander Kupco, Juergen Schmoll, Vladimir Sacek, John Leader, Daniel Schechter, Gary Beal, Robert Katz, Colin Shepherd, Es Reid, Richard McKim, William Thornton, Leonard Marek, Clint Whittmann, Martin Mobberley, and Mike Carman.

Special thanks are extended to John Watson for enthusiastically supporting the project. I would also like to thank the editorial team at Springer including Maury Solomon and Megan Ernst. Finally, I would like to thank my wife, Lorna, and sons Oscar and Douglas for putting up with my sometimes long absences from the routines of family life. I promise to redress the balance in the coming months!

Contents

Preface
The Appeal
of Yesteryear

What has been is what will be,
and what has been done is what will be done;
there is nothing new under the sun.
Is there a thing of which it is said,
'See, this is new'?
It has already been, in the ages before us.
                                        –Ecclesiastes 1:9-10

What a wonderful thing a telescope is! By altering the path of light, using lenses, mirrors, or a combination of both, this awe-inspiring construct of the human mind can let you embark on a journey across hundreds and thousands of light years of space, to witness celestial glories utterly beyond the reach of naked human eyes. How empowering it is to be able to glimpse details of our neighbor, the Moon, or the far distant planets from the comfort of one's own backyard. Telescopes are time machines, behoving us to contemplate the unfathomable natural beauty of the sky.

Telescopes are not mere inanimate objects either. They have personalities all of their own. Uncanny is the person who can't sit behind the eyepiece of a great, old telescope and not be moved by the experience, almost as if one were connecting with some deeply significant moment in the past, when curious minds observed things, perhaps even for the first time.

Over four centuries of time, this revolutionary instrument has evolved into a veritable pantheon of forms that bring the celestial realm down to Earth. They are as much part of our modern civilization as great literature is. After all, they help define humanity's soaring spirit and indefatigable curiosity for the world around us. And while contemporary telescopes continue to deliver the goods, it pays to remember that there really is nothing new under the Sun. Who can inform this author of a single ground breaking discovery, an atmospheric feature on a distant planet perhaps, or maybe a lunar feature, double star or nebulous patch per chance, that was not seen (or could have seen) and noted by our telescopic ancestors? Necromancy and nostalgia, while certainly contributing to the allure of old telescopes, certainly can't explain why they performed so well. The truth, as we shall see, is that many instruments made decades and centuries ago are every bit as good (and in some cases even better) than do many mass-produced telescopes on the market today.

One of the great psychological charms of owning a classic 'scope is that, more often than not, they were hand-built by famous makers or their highly trained technicians. The owner has a direct link to the masters of the past, which, sadly, is not seen too often in the contemporary market with its emphasis on mass production. They are one off, bespoken items, forged from the sweat and blood of optical giants.

As we will discover, many telescopes used by astronomers of generations past were broadly the equal of those employed by our contemporaries. The historical record is clear in this respect, as we shall stumble upon while recounting the extraordinary allegory of the telescope makers from the days of yore. There is much ground to cover and the book, naturally enough, had to be fairly selective in the range of artisans discussed. A classic is best described as a perfectly recognizable form, or archetype if you will, that meets all the specifications of its genre. It usually represents something of lasting worth or with a timeless quality, expressing either its sentimental or objective value at auction. It might also embody the essence of an age or help bring to life fond memories of yesteryear. And while many of these antiquated telescopes command hefty price tags, especially where provenance can be verified, it is simply not true that a classic telescope need necessarily be expensive. One need only note the extraordinary resurgence in interest in the humble 60 mm refractor across the astronomy world to see the truth in this sentiment.

In this book, we shall explore the rich lore of telescopes past, from the small and personal spyglasses of Dollond to the great observatory behemoths designed by Alvan Clark & Sons, USA; Thomas Cooke & Sons, England; and Carl Zeiss of Germany. We will unveil the extraordinary success of Japanese optical firms in the early post–World War II era, where her opticians churned out objective lenses of superlative quality that found their way into cherished brands such as Unitron, Royal Astro, Goto and Swift, to name but a few.

The book will also chart the rise of the reflector telescope from its humble beginnings in Sir Isaac Newton's study at Cambridge, through to the construction of the first parabolic mirrors that enabled celebrated observers such as Sir William and Sir John Herschel to make such enormous leaps forward in our knowledge of the heavens and our place within it. We will recount the development of new technology that did away with heavy and cumbersome metal mirrors and their replacement with silvered glass substrates. Accordingly, we shall take a look at some of history's great mirror makers, including John Calver, John A. Brashear, and more recently, the late Tom Cave, as well as some celebrated Newtonian manufacturers, including Edmund Scientific and Criterion.

The twentieth century also saw great innovations in compact telescope designs, including the Maksutov- and Schmidt-Cassegrain telescopes. This investment in new technologies, particularly the marriage of electronics and optics, led directly to the extraordinary success of companies such as Questar, Celestron, and Meade.

The refracting telescope, in particular, has enjoyed a long and distinguished history among amateur and professional astronomers, with the simple crown and flint objective prescriptions serving their needs for centuries. That said, the secondary spectrum (false color) thrown up by achromatic object glasses impelled opticians to find new glass combinations, with improved color correction. But contrary to what most contemporary amateurs believe, that search had its origins in the eighteenth century, and by the end of the nineteenth century, real progress had been made in the workshops of Zeiss, Germany, and T. Cooke & Sons of York, England.

Interest in designing color free or apochromatic refractors waned a little throughout the first half of the twentieth century but gained momentum again in the 1970s when Japanese opticians, most notably those working for Takahashi, took up the gauntlet once again, bringing to market exciting new high-performance 3-inch refractors. This was followed in the early 1980s by innovators in the United States, including Fred Mrozek and Roland Christen, who designed a new range of oil-spaced triplet apochromats for the discerning amateur astronomer.

As well as describing fully functioning telescopes from memory lane, we shall also explore some restoration projects along the way, including the refurbishment of two of Sir Patrick Moore's most used telescopes – a fine 3-inch F/12 Broadhurst Clarkson, which he purchased as a young lad, as well as a larger 5-inch f/12 Cooke refractor – arguably Moore's most used telescope back in the day.

Finally, the antique telescope market will be discussed with a view to identifying realistic expectations and potential pitfalls of prospective investors. How important is provenance? Will replacing a mirror or lens increase the value of your antique 'scope? These and other questions will be answered as we draw the book to a close. So, in the meantime, pull up a chair and settle down to read about some of the most talked about telescopes in history and something of the personalities that made them.

Yours classically,

Neil English

About the Author

Self-confessed classicist, Dr. Neil English, is the author of two influential books on commercial telescopes, both published by Springer, including the highly lauded *Choosing and Using a Refracting Telescope* and *Choosing and Using a Dobsonian Telescope*. English has conducted extensive research on the properties of the classical refractor and published widely on his findings both online and in commissioned articles for leading amateur astronomy periodicals, including *Astronomy Now*. He lives under the dark skies of rural central Scotland with his wife and two sons.

Chapter 1

The Dollond Century

In the early spring of 2010, this author was contacted by a lady who had the good fortune of inheriting an antique telescope from her late uncle. He was a bachelor and apparently a bit of a misanthrope. The lady had no idea how to set up the telescope or indeed, or what kind of condition the instrument was in. She kindly agreed to lend it to me in order that I might assess its condition. It was a 3-in. F/15 Dollond "The Student's" refractor. It came in a solid mahogany box with several eyepieces (all with solar filters built in). The lens was uncoated but absolutely pristine. It also came complete with a fabulous, full-height mahogany tripod. The tube presented in what appeared to be a green powder coat and a chrome draw tube (Fig. 1.1).

The workmanship on the instrument was quite simply in a different league to anything seen in the modern era. The telescope slotted into a cradle, using two elegantly designed clamps that required no tools. The tripod was about 5′9″ tall when the legs were folded in. The mount head itself was fashioned from some sort of bronze alloy. Locked in place, the Dollond moved with graceful elegance astride its mount, effortlessly moving from one corner of the sky to the other (Fig. 1.2).

The instrument star tested well, with nice, evenly spaced Fresnel rings seen inside and outside focus. It appeared well corrected for astigmatism, coma and spherical aberration. Bright stars like Vega and Capella reduced to hard, round disks at focus, with a faint halo of purplish light encircling them. The Cytherean phase was a delicate crescent, intensely white and sharp, surrounded by the most gorgeous halo of unfocused blue light. The telescope was tested out on a near opposition Mars over a few nights using a 5 mm ocular, presenting up well resolved views of the northern polar cap and some of the more prominent darker markings such as the *Syrtis Major*. The age of the instrument was uncertain, but after conferring with a few knowledgeable antique telescope collectors, a c.1905 date seemed plausible (Fig. 1.3).

N. English, *Classic Telescopes: A Guide to Collecting, Restoring, and Using Telescopes of Yesteryear*, Patrick Moore's Practical Astronomy Series, DOI 10.1007/978-1-4614-4424-4_1, © Springer Science+Business Media New York 2013

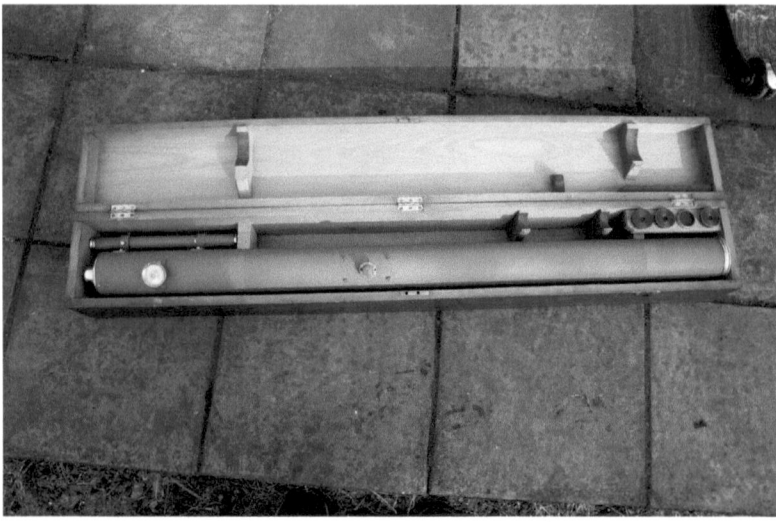

Fig. 1.1 The Dollond arrived in a beautifully made mahogany box (Image by the author)

Fig. 1.2 The alt-azimuth head of the Dollond Student's refractor

The Student's was turned on a number of deep sky objects. The Crab Nebula (M1) in Taurus delivered up its ghostly secrets in a low power eyepiece, and its crustacean morphology was clearly discerned. Higher powers rendered the Ring Nebula (M57) in Lyra as good as any 3-in. refractor ought to; a tiny smoke ring set

Fig. 1.3 Several eyepieces attended the telescope with solar filters attached (Image by the author)

Fig. 1.4 What's in a name? The all-important Dollond logo (Image by the author)

against an anthracite sky. Over in the northwest, the 3° extent of the great spiral in Andromeda (M31) could be traced out with the little Dollond. And high overhead, the glorious Double Cluster (C14) in Perseus resolved to dozens of faint stellar pinpoints (Fig. 1.4).

Fig. 1.5 The uncoated lens was in pristine shape even after a century (Image by the author)

The Student's Telescope performed handsomely on double stars, too. Almach (Gamma Andromedae), Izar and the famous Epsilon 1 and 2 Lyrae were beautifully resolved in this telescope at powers of 100× or so. Indeed, there was little to distinguish it from the images that one can enjoy with more contemporaneous instruments of the same aperture. Needless to say, it was a joy and privilege to have made its acquaintance, to have spied the starry heavens with it. But all good things come to an end, and the instrument had to be returned to its rightful owner (Figs. 1.5 and 1.6).

Yet, this delightful Dollond was but one in a long line of instruments that found their way across land and sea to the far reaches of the British Empire. At auction, it would raise perhaps $2,000 or more. Most of its wealth lies with the name engraved on the optical tube; a name that transcends national boundaries and spans the centuries.

Like many of the optical greats, John Dollond had a famously undistinguished origin. His father was a Huguenot refugee and a silk-weaver to trade, who took up residence in Spitalfields, London. It was here that John Dollond was born on June 21 1706. As a young boy, he was immersed in all aspects of the silk industry, and prospered enough to pursue a classical education, mastering ancient Greek, Latin, geometry, navigational science and astronomy. He was keen to give his son Peter the same background and encouraged him to pursue his own business interests. By the age of 20, Peter had established a small workshop making and repairing optical instruments. It was apparently a great success, as his father wrapped up his silk weaving business within 2 years to join his son.

Fig. 1.6 Ready to go. Set up takes just a few minutes (Image by the author)

The telescopes of the early eighteenth century were almost invariably of the long focus, non-achromatic variety. They were made using a single convex lens with an exceedingly gentle curvature so as to produce a very long focal length that had the effect of reducing some of the many optical flaws inherent to its design. Small, non achromatic refractors could be kept manageably short, of the order of 10–20 ft. As opticians learned to grind still larger lenses, the focal lengths grew almost impossibly long. But that didn't deter the astronomical pioneers of the day. And they came from all walks of life.

Fig. 1.7 A drawing from Hevelius' *Selenographia* displays the cratered surface of the Moon and its libration

One of the first individuals to build really long refractors was the wealthy Danish brewer turned astronomer Johannes Hevelius (1611–87) of Danzig, whose instruments reached 150 ft in length. By 1647 Hevelius published his first work, the *Selenographia*, in which he presented detailed drawings of the Moon's phases and identified up to 250 new lunar features. The *Selenographia* influenced many of the great scientists of the emerging Europe, not least of which were the brothers Constantine and Christian Huygens in Holland (Fig. 1.7).

Disillusioned by the shoddy performance of the toy-like Keplerian and Galilean spyglasses offered for sale by merchants, they set to work grinding and polishing their own lenses for the purposes of extending the work initiated by Hevelius. Between 1655 and 1659, they produced telescopes of 12 ft, 23 ft, and finally an instrument of 123 ft focal length. Instead of using a long wooden tube to house the optics, as Hevelius had done, the brothers Huygens placed the objective lens in a short iron tube and set it high upon a pole. Then, using a system of pulleys and levers, the eyepiece was yanked into perfect alignment with the objective. Christiaan Huygens used a more modest instrument (with a 2.3-in. objective and 23 ft focal length), delivering a power of 50 diameters, to elucidate the true nature of Saturn's ring system, as well as its largest and brightest satellite Titan.

Huygens not only built long refractors, he was an innovator as well. Not satisfied by the standard single convex lens that formed the eyepiece of all refractors of the day, Huygens designed a much better prototype, consisting of two thin convex elements with a front field lens having a focal length some three times that of the eye lens. The result was an eyepiece – the Huygenian – which yielded sharper images and slightly less chromatic aberration over a wider field of view than any eyepiece coming before. Curiously, Huygens also hit on the idea of lightly smoking the glass

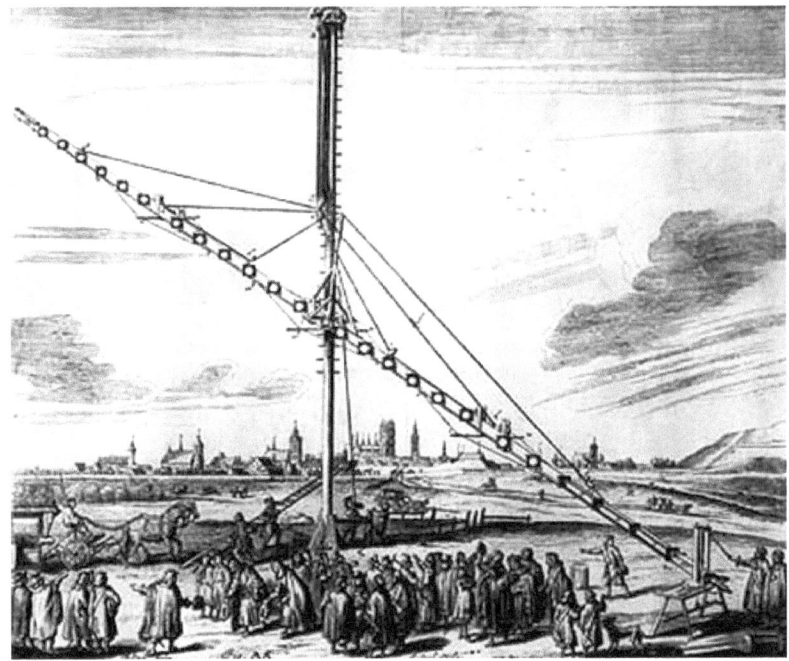

Fig. 1.8 The original 'Hevelius,' one of the largest (150-ft focus) aerial telescopes ever constructed

from which his eyepiece lenses were fashioned, so as to impart to them a yellowish tint. This cunning trick further suppressed chromatic aberration, much in the same way as a light yellow filter does when attached to a modern achromatic refractor. Huygens also appreciated the benefits of proper baffling in designing his telescopes. Placing circular stops along the main tube, these prevented stray light reflected from the sides of the tubes from entering the eyepiece, thereby increasing contrast in the image. Constantine and Christiaan Huygens produced some monster lenses, too. The largest recorded had an aperture of 8.75 in. with a focal length of 210 ft!

But even Huygens' largest telescope dwarfed into insignificance compared to the aerial telescopes made by other determined souls, such as the Frenchman Adrien Auzout (1622–91), who made telescopes with preposterously long focal lengths between 300 and 600 ft (90–180 m). Indeed Auzout also proposed the design of a leviathan telescope some 1,000 ft in length in order that he might observe the animals that inhabit the Moon! (Fig. 1.8)

The great aerial telescopes of the late seventeenth and early eighteenth century, had to be made with extraordinary focal lengths to suppress the aberrations that arise from using a single lensed objective. A biconvex lens cannot focus all the colors of the white light at a single locus. In addition, errors in figuring the lens resulted in the introduction of a suite of other geometrical, or Seidel, errors, including spherical aberration, coma, astigmatism, distortion and field curvature. The only practical way to reduce their impact on the defining power of the image was to increase the radius of curvature of the lens surfaces and that invariably meant increasing the focal length of the singlet lenses.

In his *Opticks*, Isaac Newton tested whether chromatic aberration in telescopes could be corrected by combining two lenses with different indices of refraction. The chromatic aberration of one lens would have to cancel out the other to recreate white light. But according to Newton, this could occur only if the emergent light was parallel to the incident light. Hence, quite famously, Newton abandoned the refracting telescope for the reflecting kind. Newton's erroneous pronouncement that net refraction is always accompanied by dispersion greatly impeded research into achromatic lenses for three decades following.

However, sometime between 1729 and 1733, a London barrister and amateur optician named Chester Moor Hall had serendipitously discovered that one could partially overcome chromatic aberration by combining glasses of opposite powers – a convex lens made from crown glass and a concave element made from flint. It is said that he had studied the problem for several years, firm in his belief (erroneous as it turned out) that the achromatic nature of the human eye would provide the secret design for a new type of lens. Such a doublet was able to correct chromatic aberration for the red and violet rays.

Hall intended to keep his work on the achromatic lenses under lock and key and contracted the manufacture of the crown and flint lenses to two different opticians, Edward Scarlett and James Mann. By a curious twist of fate, they both, in turn, sub-contracted the work to the same person, George Bass, who soon realized the two components glasses were being made for the same client and, after fitting the two parts together, noted the achromatic properties. Bass is rumored to have even sold a number of these instruments to private collectors. Hall failed to appreciate the importance of his invention, and it remained known to only a few opticians for another decade or more.

In 1747, the Swiss mathematician Leonhard Euler (1707–83) took up the problem and published a paper in which he showed that chromatic aberration could be corrected by sandwiching water between two concave lenses. Using the physiology of the eye as his model, Euler used mathematics to prove his conjecture. Bizarrely, he did not seek to establish the result experimentally. In Euler's world view, mathematics could always dispense with experimental science, which was a grave mistake as the events of the following years were to unravel.

When that paper reached John Dollond, then London's leading optician, he was most unimpressed. A staunch defender of Newtonian optics, he denounced it as "destitute of support from either reason or experiment." Indeed, his scorn for the famous mathematician's work went further. Mr. Euler's theorem, he said, "is entirely founded upon a new law of refraction of his own…"

Meanwhile, Euler's paper caught the attention of the Swedish mathematician Samuel Klingenstierna, who reconstructed Newton's experiments in optics and concluded that Newton's results only applied to prisms with small apex angles and that the impossibility of constructing an achromatic lens was fallacious. In 1755

Dollond, who corresponded with Klingenstierna, began his own set of experiments. In a work entitled *Experiments Concerning the Different Refrangibility of Light*, he managed to demonstrate that Newton was indeed in error:

> I cemented together two plates of glass at their edges, so as to form a prismatic or wedge-like vessel, when stopped at the edge or bases, and its edge being turned downwards, I placed therein a glass prism with one of its edges upwards, and filled up the vacancy with clear water....As I found the water to refract more or less than the glass prism, I diminished or increased the angle between the glass plates, till I found the two contrary refractions to be equal; which I discovered by viewing an object through this double prism... Now according to the prevailing opinion the object the object should have appeared through this double prism quite of its Natural colour...but the experiment fully proved the fallacy of this received opinion, by showing that divergency of the light by the prism to be almost double of that by the water; for the object, though not at all refracted, was yet as much infected with prismatic colours, as if it had seen through a glass wedge only, whose refracting angles was near 30 degrees.

It is widely assumed in the literature that Newton arrived at a 'counsel of despair' in trying to see a way forward for the singlet refractor, that manifested themselves in the aerial telescopes used by the best astronomers across Europe. "I do not yet see any other means of improving Telescopes by Refractions alone," Newton remarked, "than that of increasing their lengths."

Yet, according to the historian Rupert Hall, it would be disingenuous to think that Newton believed that it was impossible for dispersion to always accompany refraction. In a letter to the Royal Society dating from 1672, but which also appears in his *Opticks* published 20 years later, Newton described how spherical aberration might be corrected by using a double glass lens enclosing water. Hall surmises that Newton did not think it beyond the bounds of possibility that such a 'compound lens' might also correct for chromatic aberration. That said, plainly, Newton never appreciated the fact that different kinds of glass possess different refractive indices, so that his water lens was needless.

In the late 1750s, Bass mentioned Hall's lenses to John Dollond. It was the gift he was waiting for. For he immediately understood their potential and was able to reproduce their design. He applied for a patent and received one. Soon thereafter, Dollond had succeeded in creating an achromatic telescope, with a focus of 5 ft, which was presented to the Royal Society in 1758. Nefariously, though, Dollond never gave mention to Moor Hall, Euler or Klingenstierna for that matter, in his public addresses on the subject. Nonetheless, Dollond received the Copley Medal for his work, the highest scientific recognition of the time and soon thereafter became a member of the Royal Society.

The achromatic objectives consisted of two lenses made with different types of glass – crown (relatively common) and flint (with a higher refractive index), the production of which was much more difficult than anything that had come before. Flint glass, for example, was produced in small quantities only by English glassmakers who, enjoying their glorious monopoly, supplied British telescope makers with the best parts, while selling their rejects, some with considerable imperfections, to the continent. Indeed, as we shall see in the next chapter, it was not until the dawn of the nineteenth century that good flint glass began to be produced in France and Germany.

Dollond's fame reached far and wide, and everyone it seemed wanted a piece of the action. His close relations with the glass makers ensured he could select the choicest pieces of flint glass, thus ensuring the production of the finest quality objectives on the market. Even after John Dollond had patented the manufacturing process of his objectives, the company did not act particularly aggressively towards the manufacturers who soon began to copy them.

His eldest son John (1730–1821) was more vociferous, though, taking out a court order against these producers, which he won with considerable compensation. Some of his competitors subsequently found themselves in serious financial difficulty after they were forced to pay up. During the proceedings, however, the debt accrued by Dollond & Sons was beginning to become known by other scholars in optics as well as the producers of optical instruments. Indeed, a petition was drawn up by some 35 opticians in London in 1764 for the annulment of John Dollond's patent, alleging that he was not the original inventor but had knowledge of Chester Moor Hall's prior work. George Bass's name, curiously enough, is to be found on that list.

The petition had no teeth, however, perhaps because of the outcome of a court case that took place the following year in which Peter Dollond received a judgment that took the wind out of the sails of his many antagonists. In the now famous words of the presiding judge, Lord Camden: "It is not the person who locked up his invention in his scrutoire that ought to profit by a patent for such invention, but he who brought it forth for the benefit of the public."

Intriguingly, Moor Hall himself never legally contested Dollond's patent. Indeed, had he turned up in court with his fighting spirits up, he might well have brokered a better deal, but his personal circumstances were different from Dollond's. Chester Moor Hall was a wealthy and respected lawyer, a bencher of the Inner Temple and so less concerned about cashing in on the invention, in contrast to 'feral' Dollond who, as we have seen, was a dog-eat-dog businessman.

When the patent license had run its course in 1774, the commercial value of his telescopes was reduced by half almost overnight. Not that it affected the business' prosperity all that much. By this time, the name of Dollond & Sons had evolved into a truly global brand that, erroneously or otherwise, became synonymous with the finest optics in the world. Even if wasn't true. Indeed, modern investigations into some of the lesser products sold by Dollond & Sons have revealed that they could be decidedly poor compared to the quality that went into their choicest instruments (Fig. 1.9).

Nor did they exclusively market refracting telescopes. Several fine Dollond Cassegrain reflectors were made and sold by the company (Fig. 1.10).

The Reverend William Rutter Dawes (1799–1868) had been interested in astronomy as a boy, and while at Liverpool he often observed the stars through an open window with a small but excellent refracting telescope. This refractor aroused his interest in double stars, and at Ormskirk he constructed an observatory housing a 3.8-in. Dollond refractor of 5-ft focus, which he used to make careful micrometrical measurements of double stars. His measures of 121 double stars made in the period 1830–1833 were published in 1835, and a further 100 double stars measurements were added in the period 1834–1839. These were dutifully published in 1851.

Fig. 1.9 A beautiful tabletop Cassegrain designed by Dollond & Co (Image credit: Richard Day)

Fig. 1.10 The eyepiece end of a Dollond Cassegrain reflector (Image credit: Richard Day)

Chronic ill health forced Dawes to give up his pastoral work, and in 1839 he left Ormskirk to take charge of George Bishop's observatory in Regent's Park. There he continued to use his fine Dollond achromat to devote himself to the study of binary stars, and his measurements of about 250 such stars were published in 1852

Fig. 1.11 Grandfather of spotting 'scopes: a Dollond terrestrial glass (Image credit: Richard Day)

in Bishop's *Astronomical Observations at South Villa*. His results included the detection of orbital motion in Hydrae and the faint, third component of the majestic γ Andromedae (Fig. 1.11).

John Dollond was apparently one of the first opticians to use a triplet objective to correct for both chromatic and spherical aberrations, but he never published his results. Evidently, in the late 1750s, he found that by placing a concave flint glass lens between two convex crown glass lenses, the so-called triplet objective canceled out a little more chromatic aberration but significantly improved spherical aberration. It is difficult to objectively assess the putative improvements made to the Dollond refractors. That said, H. C. King, in his *History of the Telescope*, informs us that Peter Dollond constructed an instrument of 3.75 in. f/11 specification. According to one source, it rendered an image "distinctly bright and free from colors when charged with a magnification of 150." Indeed, the report goes on to state that the same instrument was capable of holding powers of 350× 'without breaking down.' The triplet was used to resolve fairly tricky double stars, such as Epsilon- and Iota Bootis, Gamma Leonis and Eta Coronae Borealis (which at the time of writing had a 0.9″ separation).

Indeed, amateur astronomer Thomas Jensen, based in Bornholm, Denmark, stated that a late friend of his had once tested a Dollond 3.5 in. f/12 triplet achromat from c. 1775, and it performed almost as well as his own Zeiss AS100/1000, save for a slightly dimmer and off color image, owing to the greenish crown glass it employed. Color correction was almost as good as his Zeiss, and spherical correction was spot on. "It was incredibly sharp!" he says, "and as a lunar, planetary and double star scope it did very, very well and left little to be desired."

Alas, we know next to nothing concerning the methodology Peter Dollond employed in fashioning his lenses, although Jean Bernoulli (Jr.), who visited Dollond's workshops in 1769, came away with the distinct impression that Peter did not possess his father's theoretical knowledge. Indeed, he claimed that the junior Dollond worked largely by trial and error, testing systems of lenses to see which ones worked best and discarding those that were less than satisfactory. What is more, Bernoulli was not particularly impressed with his larger instruments either:

> *You make a great mistake when you imagine that an astronomical instrument bearing the name of Dollond must be excellent in every feature. If you receive one which has this quality, it is a sign that it has not been finished by one of Dollond's workmen; often to maintain his reputation he has the mountings and divisions made for him by his brother-in-law, M. Ramsden, who passes for one of the best artists for this work in London.*

Bernoulli was speaking, of course, of a one Mssr. Jesse Ramsden (1735–1800), who rapidly ascended the stairway of notoriety as an instrument maker before enjoining himself to the Dollond dynasty, by marrying John Dollond's daughter Sarah. As part of the dowry, Ramsden received a share in the patent for manufacturing achromatic lenses. It was a move that benefitted everyone, for Ramsden flourished in a workshop of his own design, using objectives produced by Dollond to construct fine refracting telescopes. Arguably his most celebrated work was a 5-ft vertical circle, which was finished in 1789 and used by Father Giussepe Piazzi at Palermo Observatory, Sicily, to compile a marvelous new catalogue of some 8,000 stellar positions as well as attempting to measure the parallax of several bright stars. Indeed, so fine were Ramsden's technical achievements that he, like John Dollond before him, was elected Fellow of the Royal Society in 1786 followed in 1795 by the award of the Copley Medal.

In 1766, Peter moved the business to more opulent premises at 59 St. Paul's Churchyard, where he was joined in 1768 by his older brother, John, who became a partner in business and was shortly thereafter appointed optician to King George III and the Duke of York. In 1783, the Dollonds began to provide their telescopes with brass draw tubes and developed their characteristic brass-bound mahogany tubes. Complementing these tubes, the Dollonds offered them for sale coupled to a wide range of sturdy mounts. Small telescopes could be placed on elegant, tabletop mounts of the folding claw and pillar variety, which proved very popular with both the upper and professional classes.

Many Dollond instruments were carried off to distant lands and pressed into service to observe the transits of Venus that occurred in 1761 and 1769. The idea of observing the transit from different places came from Sir Edmund Halley, the second Astronomer Royal. In 1716, he suggested that if viewed simultaneously from different points on the globe, the transit could be used to determine the Earth-Sun distance and thus, using Kepler's laws of planetary motion, deduce the size of the Solar System (Fig. 1.12).

In 1761 the Royal Society dispatched Nevil Maskelyne (later the fifth Astronomer Royal) and Robert Waddington to the island of St. Helena, while Charles Mason and Jeremiah Dixon traveled to the Cape of Good Hope.

Fig. 1.12 A Portable Dollond telescope similar to that used by Cook to observe the transit of Venus in 1769 (Image credit: National Maritime Museum)

Unfortunately, bad weather ruined the observations of Maskelyne and Waddington, so a comparison between the two observations could not be made.

In 1769 the Royal Society funded and organized a new set of expeditions. This time, astronomers were sent to Hudson Bay, Cornwall, Ireland, Norway and Tahiti the southern Pacific Ocean, to observe the second transit of Venus that century. Lieutenant James Cook was chosen to command the expedition to Tahiti.

Dollond's business success was greatly aided by the needs of the army and navy during the Napoleonic Wars. The enormous demand for his instruments by the British army and navy led Dollond to develop their characteristic 'Army telescope,' adorned by a mahogany bound body and brass collapsible tubes. These 'signaling telescopes,' as they came to be known, ranged in length from 14 to 52 in.. The company also developed their so-called 'pocket perspective glasses' for the general public (Fig. 1.13).

Telescopes, of course, were just a small part of an ultimately hugely successful business. Dollond's achromatic lenses found their way into microscopes and a rich array of surveying instruments, including theodolites, sextants, octants and spectacles. Dollond wares, great and small, quickly donned the iconic mantle of the Age

Fig. 1.13 Dollond marine telescopes from c. 1800 can be purchased for just a few hundred dollars (Image credit: Fleaglass.com)

of Enlightenment, reaching across the vast British Empire and beyond. In Austria, Leopold Mozart, the father of the greatest musical prodigy the world has ever seen, set up a small 3-in. Dollond refractor in his garden while his gifted children played the harpsichord from their secure surroundings. And during the battle of Copenhagen in 1801, Admiral Nelson famously placed his blind eye to a Dollond signaling spyglass and said to his Flag lieutenant, "You know Foley, I have only one eye. I have a right to be blind sometimes. I really do not see the signal."

Perhaps the most lucrative contract secured by the Dollonds derived from that which was brokered with the Hanoverian kings of England, who commissioned many instruments from Dollond & Co. over the decades. Dollond telescopes were also in great demand by the emerging Patrician families of the newly founded United States. For example, while visiting London, the polymath and future president, Thomas Jefferson (1743–1826), purchased his first achromatic telescope, of the latest triplet design, from the Dollond brothers in 1786.

Researching the archives shows that Jefferson acquired a second Dollond telescope in 1793. Both instruments survive to the present day. However, in the same year, Jefferson acquired the most valuable instrument in his collection, an equatorially mounted telescope (with optics fashioned by Dollond) by Jesse Ramsden, whom he considered preeminent among makers. With "this noble instrument," as he called it, he fixed the meridian at Monticello and viewed the solar eclipse of 1811. Judged by one scholar to be "unquestionably the most sophisticated instrument in the United States" at the time, Jefferson's equatorial refractor provided the cornerstone of his work in determining a reliable method for calculating longitude by lunar distances without the use of a timepiece.

Ironically, although he often expressed a desire for a more powerful telescope, Jefferson apparently never acquired one capable of viewing the eclipses of Jupiter's satellites, a necessity for the other common method of determining longitude. Indeed, the telescopes he acquired were more akin to the retractable draw tube variety than a 'proper' large aperture instrument that could undertake serious scientific investigations.

Peter Dollond stepped down from directing of the optical works at 59 St. Paul's Churchyard in his 87th year, retiring to his stately residence at Richmond Hill. In early 1820, he moved to a new home in Kensington but died just a few days after moving in. Thereafter, the business passed to his nephew, a one George Huggins (1774–1852), who later changed his name to George Dollond. It was an auspicious game changer for the young man, who inherited both his uncle's mechanical skill and his grandfather's sound grasp of optical theory.

Despite the difficulty of obtaining optical glass blanks greater than 4 in. in diameter, a number of larger Dollond refractors have been documented in the literature. Most were made under George's watch, and so date to the first half of the nineteenth century. The largest observatory-class refractor in the United States before 1830 was a 5-in. Dollond achromat, which was acquired by Yale University in 1829. It was bequeathed to the university as a gift from Sheldon Clark and was housed in Yale's astronomical observatory, called the *Atheneum*, which, at the time, consisted of just a spacious room perched atop a large tower on the university campus.

The telescope was rather poorly mounted, however. It was erected on castors and moved from window to window. Moreover, it could not reach altitudes much over 30°, severely limiting its efficacy. Yet, in acquiring the 5-in., the university achieved the enviable distinction, for a while at least, of having the largest refractor in America. It was put to good use monitoring the motions of Jupiter's large Galilean moons, and in 1835, Professor Dennison Olmsted and Tutor Elias Loomis used the same telescope to spot Halley's Comet upon its expected return in the August of that year. The *Atheneum* was eventually demolished in 1893 and the Dollond telescope donated to the Smithsonian Institute in Washington, D. C., where it can still be seen today.

Arguably one of Dollond's finest instruments was his 5-in. F/24 refractor, which was, for a short time, the largest of its kind ever constructed and was erected at the Royal Observatory, Greenwich, for observations of the eclipses of Jupiter's satellites and lunar occultations of faint stars. The archives also attest to an observatory

Fig. 1.14 George Dollond continued to secure the patronage of the British Crown as Instrument Makers to her Majesty the Queen during the early to mid-Victorian Era (Image by the author)

built by John Dillwyn Llewelyn (1810–1882) in 1851 on his estate at Penlle'rgaer near Swansea. It housed a 4.75-in. Dollond refracting telescope on an equatorial mount under a rotating cylindrical roof.

The stone building included a private laboratory. Llewelyn had an interest in photography and was an early pioneer in the field. He and his daughter Thereza Llewelyn (1834–1926; later Thereza Story-Maskelyne) used the telescope in the mid-1850s to take one of the earliest daguerreotypes of the Moon. The building still stands on the Penlle'rgaer estate grounds, and some renovations were carried out in 1981 to safeguard the structure. An even larger Dollond refractor (7-in. aperture) was reputedly housed at Mr. Bishop's observatory, South Villa, Regent's Park, London, and apparently saw active service between 1836 and 1861, before being translocated to Twickenham and finally dismantled in 1874. It has a 7-in. refractor by Dollond, with which a certain Mr. Hind discovered ten minor planets and several comets, as well as constructing a detailed map of stars near the ecliptic.

George Dollond built numerous precision astronomical instruments with great attention to detail. He also invented an "atmospheric recorder" by which continuous measurements of temperature, wind, rainfall, humidity, pressure and other weather data were printed on rolls of paper. After Peter Dollond died in 1820, George Dollond ran the family business until his own death, on May 13, 1852, which marked the end of a remarkable Dollond century (Fig. 1.14).

The business continued to be successfully administered by members of the family. But by the end of the nineteenth century, fierce competition from a number of telescope manufacturers, both at home and overseas, forced the company to re-evaluate its future. As a result, the company ceased producing telescopes and concentrated on the lucrative spectacle trade. (After all, Peter Dollond had invented bifocal spectacles back in 1781!) In 1927 Dollond & Co merged with Aitchison & Co, forming Dollond & Aitchison, the High Street chain of opticians still in business today.

The end of the Dollond century did not signal the death knell of British optical excellence. Indeed, as we shall see in the next chapter, we'll explore the extraordinary allegory of another Englishman who arose from obscurity to create some of the largest and finest telescopes the world had ever seen.

Chapter 2

A Yorkshireman Makes Good

The ancient English city of York has enjoyed a long and illustrious history spanning two millennia. Founded by the Roman governor of Britain, Quintus Petillius Cerialis, in A. D. 71, it lies at the junction of two great rivers – the Ouse and Foss – and quickly grew from a garrison town into a major northern city of the Roman Empire. The second century Spanish emperors, Trajan and Hadrian, knew the place. The third century African *Princeps*, Septimius Severus, died there, and in the fourth century, Constantine the Great was proclaimed the western Augustus by his troops within its walls.

Fortified and expanded by the Vikings and Normans who followed them, York also basked in the noon day brightness of the Industrial Revolution, attracting all manner of skilled artisans to its bustling streets. And it is here that our story begins, when and where a young man named Thomas Cooke founded a telescope- making dynasty that restored Britain's talent for scientific innovation throughout the Victorian Era and beyond.

Cooke was born on March 8, 1807, in Allerthorpe, Yorkshire. The son of a shoemaker, he received only the briefest of formal education, when after 2 years at an elementary school, he was put into his father's trade. But it soon became evident that such an occupation was not for the dreamy boy who pined for maritime adventures. Bright and curious, Cooke soon resumed his learning, teaching himself mathematics, navigation and astronomy. Fortunately for us hopeless telescope junkies, Cooke never did set sail on the high seas, his mother having persuaded him (insisted?) to seek local employment instead. From 1829 to 1836, he pursued a teaching career as an assistant schoolmaster and private tutor. And it was during this time that he met his future wife, Hannah Milner.

Cooke's interest in practical optics impelled him to begin work on his first telescope, one of the lenses of which he ground from the bottom of a whiskey tumbler and mounted the objective inside a tin tube that he soldered together from scrap

N. English, *Classic Telescopes: A Guide to Collecting, Restoring, and Using Telescopes of Yesteryear*, Patrick Moore's Practical Astronomy Series, DOI 10.1007/978-1-4614-4424-4_2, © Springer Science+Business Media New York 2013

metal. That same telescope was bought by a one John Philips, then Curator of the Yorkshire Museum but who later became an active member of the British Association for the Advancement of Science. Philips was to prove a powerful ally in the advancement of Cooke's subsequent career.

His marriage to Hannah was bountiful, too. She bore him seven children in all. Two of his sons, Charles (1836–98) and Thomas (1839–1919) subsequently joined him in the business he founded in 1837, at 50 Stonegate, York, with a loan of £100 from his wife's uncle. From this unassuming, rented premises, Cooke began work repairing and making instruments to order.

Throughout the eighteenth century, Britain had established a solid lead in optical glass manufacture, attributed no doubt to the extraordinary success of the Dollond dynasty and the many artisans who grew up around them. Yet, by the second decade of the nineteenth century, England's optical glass industry was crippled. In a penetrating modern analysis, historian Myles Jackson referred to the affair as "the British Crisis," in which the government maintained a stranglehold on the glass furnaces by enforcing heavy taxes on the manufacture of crown and flint glasses, while domestic types were exempt from duty. The motivations of Her Majesty's government lay with the large quantities of wood and (later) coal, consumed to produce the melts. Those raw materials – the energy resources of Empire – were to be prioritized for other purposes. Indeed, although at the beginning of the eighteenth century 13 optical glass works were in operation across the country, only three remained by 1833 and with them a drastic loss of skilled artisans. It was to be another 12 years before Parliament repealed these heavy tax levies.

In the early 1820s, though, in Bavaria, Germany, optical glass working was undergoing a bit of a Renaissance. A Swiss bell maker turned glass worker Pierre Guinand, under the aegis of Joseph Fraunhofer, hit upon a way of making larger blanks of both flint and crown using a new and improved stirring process.

Fraunhofer was born in the small Bavarian town of Straubing, Germany on March 6, 1787. The eleventh and last child of Franz Xavier Fraunhofer, a glazier to trade, the boy was plunged into misfortune from an early age when he lost his mother at 11 and his father just a year later. Thereafter, Fraunhofer was apprenticed to the mirror maker and ornamental glass cutter P.A. Weichselberger based in Munich. But after serving just 2 years of apprenticeship, disaster once again struck when Weichselberger's house collapsed. Luckily, Fraunhofer was protected by a cross-beam and escaped serious injury. Traveling to the scene of the debacle, Prince Elector Maximillian Joseph IV of Bavaria was apparently so moved by the fate dealt to the young Fraunhofer that he invited him to stay at his castle at Nymphenburg, ordering his privy councillor, a one Joseph von Utzsheneider, an influential politician and entrepreneur, to look after the youth.

In 1806 Fraunhofer was offered a junior post at the Munich Institute by Utzschneider, making fine optical instruments, where his extraordinary abilities were soon realised. Within a year, he was grinding and polishing lenses and soon after took charge of a workshop and several apprentices. Utzschneider moved his business to Benediktbeuern, where he had founded a glass melting workshop. It was here that Fraunhofer met the Swiss Pierre Louis Guinand, a specialist in melting high quality crown and flint glass. Utzschneider instructed Guinand to introduce Fraunhofer to the secrets of glass melting. After 1809, Fraunhofer was already a partner of the firm and in charge of building optical instruments: microscopes, opera glasses and astronomical telescopes. The firm produced everything in house; the optical parts, mountings, clockwork mechanisms, precision shafts, tube and the young man rapidly gained a reputation for producing achromatic doublets of excellent quality. Fraunhofer eschewed the trial and error processes used by his predecessors. He firmly understood the relationship between the refractive index of the glasses he employed, their curvature and resulting dispersive powers. These new techniques enabled Fraunhofer to design and build a giant 9.5-in. refractor, the largest aperture refractor the world had ever seen, which was installed at the Dorpat Observatory, Russia, and entrusted to F.G. Wilhelm Struve on November 10, 1824. Upon its arrival, Struve inspected the instrument and recorded his memories of its arrival for posterity:

> On opening the boxes, it was found that the land carriage of more than 3000 German miles (close to 1500 English miles), had not produced the smallest injury to the instrument, the parts of which were most excellently secured. All the bolts and stops, for instance, which served to secure the different parts, were lined or covered with velvet; and the most expensive part (the object-glass) occupied a large box itself; in the center of which it was so sustained by springs, that even a fall of the box from considerable height could not have injured it. Considering the great number of small pieces, the putting together again of the instrument seemed to be no easy task, and the difficulty was increased by the great weight of some of them; and unfortunately the maker had forgotten to send the direction for doing it. However, after some consideration of the parts, and guided by a drawing in my possession, I set to work on the 11th, and was so fortunate as to accomplish the putting up of the instrument by the 15th; and on the 16th (being a clear morning) I had the satisfaction of having the first look through it at the Moon and some double stars. I stood astonished before this beautiful instrument, undetermined which to admire most, the beauty and elegance of the workmanship in its minute parts, the propriety of its construction, the ingenious mechanism for moving it, or the incomparable optical power of the telescope and the precision with which objects are defined.

The German furnaces, unlike those in England, were still fueled by wood but didn't generate the same kind of heat as coal, rendering the homogenization of the melt more problematic unless a more effective way of stirring it were achieved. Guinand's technique involved constant stirring of the molten glass using a cylinder of fire clay, bringing bubbles to the surface and ensuring the melt was thoroughly mixed from its complete fusion until, after very slow cooling, it became too viscous to stir longer. Guinand succeeded in that goal where many others failed, employing a precise combination of time, temperature and stirring. Fraunhofer however, was a paranoid soul, and, as a result, went to great lengths to keep Guinand's pioneering new technique a closely guarded secret.

Nonetheless, a letter written by a certain Reynier of Neuchatel, Switzerland, to the Council of the newly founded Astronomical Society of London in 1821, stated that Guinand could deliver high quality optical glass blanks up to 12 in. in aperture. Intrigued, the Council invited Guinand to submit samples for inspection. The largest flint blanks were disappointingly small, just 2 in. in diameter and were given to the very capable London optician Charles Tulley (active 1780–1824), who combined it with fine English crown glass to produce a telescope that was described as "trifling" in size but excellent in performance.

Indeed, Tulley apparently received even larger flint disks from Guinand. One was a very respectable 7.25 in. in aperture which he attempted to achromatize with a similar-sized plate glass. The resulting mating was poor. Tulley then combined it with English crown glass and produced an instrument of 6.8 in. clear aperture and 12 ft focus. He then invited George Dollond and Sir John Hershel, among others, to observe Saturn, Jupiter, the Virgo 'nebulae' and a variety of difficult double stars through it. The unanimous verdict was impressive and helped to consolidate Tulley's optical reputation both at home and abroad. Perhaps one of his finest instruments – a 5.9 in. achromatic – went to the noted English double star observer, Sir James South (1785–1857), who entrusted Tulley with the task of grinding it and which South himself considered to be the finest in the world. Many choice antique instruments bearing the name of Charles Tulley are considered highly collectible today, some of which have commanded five-figure sums.

The Admirable Admiral

The early Victorian period represented an exceptionally changeable time for astronomy. On the continent, the French, Russians and Germans had established large observatories with professionals at the helm. The United States, still a sleeping giant, had not yet realised her latent talent for producing some of the finest refracting telescopes in the world. But as John Weale reported in an account of London's observatories in 1851, privately owned establishments, run by wealthy amateurs, were all the rage across England and indeed had become 'fashionable'. Immersed in this 'gentleman astronomer' culture, William Henry Smyth, a retired sea captain and later admiral in the Royal Navy, flourished.

Smyth's childhood was, by all accounts, a happy one, with long days filled with adventure and romance in equal measure. But for us amateur astronomers, it is his last forty years that we cherish most. The son of an American loyalist who had returned to Britain after the Revolution, Smyth fancied himself as a bit of a Captain Cook. After climbing on board a merchant ship that had docked in the Thames, he had run away to sea as a lad. And there he stayed, joining the Royal Navy during the height of the Napoleonic Wars, when Lord Nelson had risen to become the hero of 'Free Europe'. Much of his early naval career was spent in the Mediterranean, assigned to the pro-British naval base at the Kingdom of Naples and Sicily. By his early twenties, he had been promoted to his first command of a small squadron in the Straits of Messina, where he helped keep the anarchy of the Barbary Pirates at bay.

And it was during these years that Smyth, invited to the Court of King Ferdinand IV of Naples, met the illustrious Italian astronomer, Father Giuseppe Piazzi, who had already earned a piece of immortality by discovering the first asteroid, Ceres, back in 1800. Despite his fervent Protestantism, Smyth found a kindred spirit in the Italian Catholic who was to make a lasting impression on the upwardly mobile Englishman, by opening the young man's eyes to the possibilities a scientific career might bring. Smyth soon sought out the great observatories of Europe, learning how to use the astronomical instruments at their Royal Observatory, Greenwich, as well as those established at Palermo, Sicily.

Curiously, in the same way that Lord Nelson had met his future wife, the Lady Hamilton, while at the Neapolitan Court, so too did Captain Smyth become acquainted with Miss Annarella Warrington, the daughter of a future British Consul, at the same court. They subsequently married, and unlike Nelson's, the matrimonial union proved a long and happy one. As father to three daughters – Henrietta, Ellen and Rosetta – he cultivated their passion for civilised learning, instructing them in practical astronomy, navigation and mathematics. Their son, Charles Piazzi Smyth, was later to become one of the most notable figures in the Victorian scientific movement and indeed later became Astronomer Royal for Scotland.

After the fall of Napolean and the liberation of Europe, Smyth served out his time in the eastern Mediterranean, undertaking hydrographic surveys. Supporting a young family, he remained in Naples until 1825 but thereafter 'retired' to England, living out the life of a country Laird in Bedford, where he soon assumed the mantle of the Gentleman Astronomer, a persona that subsequent generations would hold in great affection.

From his opulent, country home, Smyth constructed an elegant observatory, equipping it with a transit instrument and an accurate timepiece with which he could measure both the right ascension and declination of stars as they trundled across the meridian. He also had in his possession a small refracting telescope astride a solid mounting, which he could move round his estate. Charging it with high-quality micrometer eyepieces, he undertook measurements that would better quantify the refractive index of the air, by accurately recording stellar positions.

Some time later, Smyth acquired a 5.9″ equatorial refractor from Sir James South. Though probably average by modern standards, it nonetheless represented one of the largest and most sophisticated refractors in Britain. With this telescope, Smyth began a long programme of original research on double and variable stars, as well as kinematic studies on the proper motions of nearby stars.

Smyth's new-found passion for advancing the cause of visual astronomy blossomed in the fertile soils of the British Empire, where his contemporaries – men of the ilk of Sir John Herschel, William Rutter Dawes and Sir James South – were carrying out exciting new researches from the comfort of their grand estates. It was during these seminal years that Smyth first made his acquaintance with the wealthy barrister and squire of the great Hartwell Estate, Aylesbury, a one Dr. John Lee. Passing through Bedford whilst travelling to the County Quarter Sessions Courts, Lee often stopped off at Smyth's house, where he enjoyed the use of the Admiral's fine instruments. Indeed, under Smyth's aegis, Lee constructed his very own observatory (Hartwell), equipping it with the finest astronomical contraptions money could buy. And though they remained firm friends, kindred astronomers as it were, their personalities couldn't have been more different.

Dr. Lee, the embodiment of Victorian idealism, was a teetotaller, eschewing the activities of gambling and the pleasures of hunting – the time-honoured pastimes of many of his peers. Smyth, on the other hand, had acquired many of the habits of his seafaring comrades, indulging in the culinary delights of good food and the various libations his social station had accustomed itself to, as well as expressing a decid-edly more conservative political worldview. Yet, each man thrived in each other's company, hosting numerous astronomical gatherings. Indeed, rumour has it that while Dr. Lee was constructing his lavish observatory, he would issue a certificate of merit to anyone who would lay a commemorative brick towards its completion. But his brick layers were no ordinary plebs; indeed they were the *nobilitas* of the *Imperium Britannicum* and future presidents of the prestigious Royal Society, including names like Airy, Brewster, Struve, Herschel and Rumker. Wealthy 'com-moners' were also welcomed with open arms, including the tycoon landowner-brewer, Samuel Whitbread, who had himself built private observatories at his pallatial London home, at Eaton Square.

Both Dr. Lee and Captain Smyth, as active Fellows of the Royal Astronomical Society (FRAS), began to publish numerous papers on various astronomical topics. But it is arguably Smyth's 1844 work, *A Cycle of Celestial Objects*, that he became more generally known. Distilling some twenty years of experience in matters of practical astronomy and astrometry, which greatly aided Smyth's rise to notoriety, even among upwardly mobile nobodies. In this beefy, two-volume work, Smyth published many useful tables (which the American poet Walt Whitman would later write about with derision) of the celestial real estate he had visited, together with invaluable advice on their location and study. It is here that one will also find a treatise on Gamma Virginis, the first double star, the orbital aspects of which had just been established through careful study.

Yet, seen through the lens of modernity, *Cycles in the Heavens* could certainly not be considered to be an easy tome to digest. Indeed, Smyth presupposes that his read-ers possess quite a sophisticated background in trigonometry, optics and linguistics. But as terse as it sometimes seems, Smyth's *magnum opus* need not be construed as being deliberately elitist.

Smyth was a man of his time, classically trained for the Age of Empire. If any-thing, it just illustrates the sheer gulf between the haves and have nots of the day, as well as the circles within which the good Admiral and his chums moved. For Smyth, the squalor of a London slum was a distant and unthinkable possibility.

In the *Cycle*, Smyth described in great detail, the constitution of his own obser-vatory at Bedford. The 'truncated dome,' ran on wooden balls, where, on a favour-able evening, his beautiful, 5.9″ Tulley refractor peered out. In addition to the main circular space, Smyth also had constructed ancillary 'transit' and 'computing' rooms, where his rough numbers, derived from the micrometer, could be reduced.

The cost of building such a grand observatory must have been prohibitive to all but the most well to do folk, and while Smyth never alludes to its cost, another Gentleman astronomer, the wealthy surgeon, William McClear does. Indeed, in 1882 McClear commissioned a local carpenter to erect a six foot diameter wooden dome, which, at £50, he deemed 'economical'. The reality however, is that this sum of money would have kept a working class family, with several dependants, sweet for an entire year!

In retrospect, we only know so much about the activities of the Victorian Grand Amateur culture because Dr. Lee's *Albums* of the Hartwell Estate have been so well preserved. Another issue that needs to be clarified is the role of women in such a society, with the common misconception that the fairer sexes were really second class citizens in Victorian society holding sway. Yet, the *Albums* clearly record the attendances of ladies and children who appear to have been warmly welcomed into these grandiose Victorian 'mancaves,' drinking up the views through the magnificent refractors erected therein, or perhaps weighing up the latest theories of cosmogeny with their spouses and other male acquaintances.

Captain Smyth had his place in the pecking order too. He was not as wealthy as Dr. Lee. Indeed, the good Admiral once described himself as a 'half pay naval officer' implying that he lacked the true, landed wealth of his magesterial friend at Hartwell. By 1853, he had acquired the rank of rear-, followed shortly afterwards by vice-admiral. Only in 1863 was Smyth promoted to full Admiral, though by that time, he was nothing more than a beached officer.

Indeed, the publication of Smyth's *Cycles* in the 1840s may have reflected an underlying financial crisis in his life. Diligent research carried out by the distinguished historian of astronomy based at Oxford University, Dr. Allan Chapman, has shown that the early 1840s were characterised by a volatile financial market, both at home and overseas, with many banks and mercantile companies crashing out. Indeed, this may well have been the motivation behind the sale of Smyth's 5.9″ refractor to Dr. Lee and its re-housing at Hartwell House.

In the autumn of his life, Admiral Smyth's conviviality was known the length and breadth of the country. We now know of many correspondences he made with gentlemen in Liverpool and Nottingham in the 1850's. Indeed, by the 1860's, there would have been few places in the British Isles that did not have an Astronomical Society of sorts, equipped with ever more impressive instruments donated by the euergetism of wealthy patrons. When Smyth started his astronomical adventures, a 5.9″ object glass was considered world class. Thirty years on, refractors as large as 10 inches were being used by gentlemen amateurs continuing in the good Admiral's footsteps.

After 70 trips round the Sun, Smyth bade farewell to active observing but lived a further seven years. His life-long friend, Dr. Lee, survived him by only a year and with their passing, much of the instrumentation became entrusted to the RAS. The famous 5.9″ equatorial is now in safe keeping at the Science Museum in South Kensington, London.

Tulley's obtaining of large glass blanks from Guinand proved to be the exception rather than the rule, though. Some of the greatest British scientists of the age tried everything to get their hands on the details of the new German technology. Traditional diplomacy quickly descended into bribery (Fraunhofer was offered £25,000 for information), but without success. Circumstances changed, though, as they invariably do. In the summer of 1826, the Mozart of practical optics died prematurely, opening the way for glass makers to sell their secrets to the highest bidder. It marked the end of Fraunhoferian hegemony and the beginning of a new age in British optical glass making.

Cooke learned of these new techniques and applied them to build his first 'serious' telescope, a 4.5-in. equatorial, for a well-to-do lawyer, William Gray. McConnell describes the process of preparing the glass in her book, *Instrument Makers to the World*:

Glass for optical purposes was not blown or rolled, but allowed to cool slowly in its pot, then removed and broken up. Flawed pieces were discarded and the remaining blocks cut into disks. These were sent to the optician who ground them into the shape of a lens with perfectly spherical surfaces. The convex shape was ground in a saucer shaped iron plate covered with pitch, hatched by cross grooves to take away the waste material. The pitch was warmed and covered with rouge – a fine abrasive. The lens was then rubbed and turned by hand or machine to achieve the desired curve.

Beyond that, little else is known about the precise techniques Cooke employed to figure his object glasses. But what we do know is that the prescription of the early Cooke objectives differed a little from the standard Fraunhofer template, with the latter usually (although there are apparently some variations) having a more strongly curving outer crown element than its Fraunhoferian counterpart. What's more, the fourth (innermost) element of the Cooke object glass was flat and so didn't require figuring.

The 4.5-in. equatorial was apparently a great success, for news of its quality spread far and wide. Cooke made more instruments and built his reputation. His second large commission was for a 7.5-in. equatorial for a Mr. Hugh Pattinson. When his friend, the noted astronomer Isaac Fletcher had a chance to evaluate its optical and mechanical performance, he was so duly impressed that he wrote to Sir George Bidell Airy, then the Astronomer Royal (the man who, quite literally, divided the world in two, by establishing the new Prime Meridian at Greenwich), suggesting that Cooke be commissioned to make the lens for the Cape Observatory in South Africa. And although that job had already been contracted out to another party, it certainly helped Cooke's reputation and finally enter the conversations of the inner sanctum of the astronomical elite (Fig. 2.1).

The Victorian Era was no Antonine Age. Indeed the *Imperium Britannicum* had not seen a year free of the ravages of war in all the days of Victoria's reign. And while the construction of large equatorially mounted refractors were statements of scientific prestige, Cooke & Sons played their dutiful part in helping to sustain the machinery of the Empire – theodolites for surveying, spyglasses for naval officers, range sights for more accurate killing machines, and magnetometers for mining geologists in their ever more efficient plunder of colonial resources. With the railway coming to York in 1839, many new opportunities appeared across the country, and Cooke's goods found their way into every major optician's premises from Sutherland in the north to Cornwall in the south. Cooke also displayed a penchant for practical horology and set his considerable mechanical abilities to good use, manufacturing turret clocks for church towers.

In 1855, Cooke moved to bigger premises, the Buckingham Works, at Bishop Hill in York, where factory methods of production were first applied to optical instruments. Employing six workmen and one apprentice, everything, with the exception of the glass blanks, were made in situ – workshops for glass, brass and wood and a foundry where all but the largest castings were made. In the same year, Cooke decided to bring his new wares to the continent, exhibiting his instruments at the Universal Exhibition in Paris – the NEAF of its day. His gamble paid off, for he came away from the event with a First Class Medal for his clock-driven equatorial mounting and a ringing endorsement from the chattering classes (Fig. 2.2).

Fig. 2.1 Thomas Cooke (1807–1868)

Cooke received commissions for several more large refractors (from 5- to 10-in. apertures), but perhaps the most prestigious of all came from an order by Prince Albert, who, in 1860, summoned Cooke to Osborne House, on the picturesque Isle of Wight, to discuss the construction of a telescope for the viewing pleasure of the royal family. They chose a fine equatorial refractor of 5.25-in. aperture.

Arguably one of Thomas Cooke's greatest achievements was the construction of the 25 in. 'Newall' refractor. Built for Robert Stirling Newall, the story behind the completion of this great telescope is a particularly somber one. After accepting the commission for the giant lens, Cooke greatly underestimated the length of time it would take to complete it. Indeed, he surmised that it would take no more than a year. What is more, in an attempt to undercut a quote from his rival, Sir Howard Grubb, he charged too little for the work.

These realities conspired in such a way that Cooke failed to meet several new deadlines he agreed to with Newall. What is more, Newall accused Cooke of 'taking his eye off the ball' as it were, and even threatened to withhold further down payments on the project. The probable reality, as McConnell convincingly argues, is that Cooke was under enormous pressure to complete other sizable commissions on time. Thomas spent his twilight years a sickly man, finally giving up the ghost on October 19, 1868, aged 61.

In his will, Cooke bequeathed 'everything' to his wife, who immediately instructed her sons to see the Newall project to completion and the telescope trans-

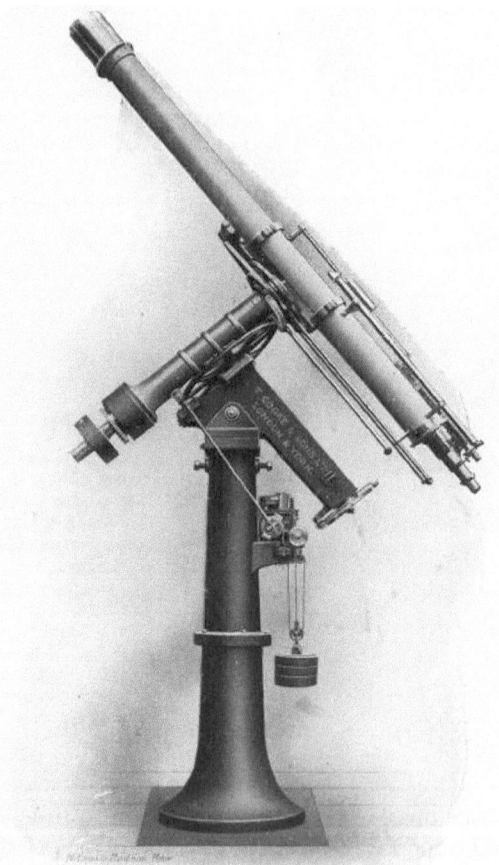

Fig. 2.2   An instrument for a well-heeled amateur c. 1899 (Image credit: Doug Daniels)

ferred to his estate in Ferndean, Northumberland. It arrived in 1870 and became fully operational a year later.

The Newall telescope enjoyed the distinction of being the largest refractor in the world for only a year, when Alvan Clark's latest monster refractor, erected at the U.S. Naval Observatory, Washington, D. C., literally inched it out of first place in 1872. After Newall's death, the instrument was donated to Cambridge University and lauded for its fine optics. By the 1950s, however, the Golden Age of Astrophysics having come and gone, the telescope fell into disuse. Finally, in November 1958, the decision was made to sell it to the Greek National Observatory, which housed it in a magnificent dome atop Penteli Mountain, just a day's walk north of Athens.

Messrs. Cooke continued to secure enough orders to keep the business (which now had grown to a workforce in excess of 100) ticking, and in the decades ahead, it continued to provide both private individuals and public observatories scattered

Fig. 2.3 A 4-in. F/15 Cooke branded A. Ross. c. 1860 (Image credit: Richard Day)

across the world with large instruments, including an 18-in. refractor for the Brazilian National Observatory.

Several optical firms in the United States and in Germany were employing new and more sophisticated lenses, using techniques that rivaled or exceeded the quality produced by the Cooke brothers. To add insult to injury, many other, smaller firms were beginning to compete with the British optical giant, undercutting their powerful rival.

T Cooke & Sons evolved and adapted to their changing circumstances as best it could. For example, the company was known to supply telescopes for re-branding, such as the fine 4-in. F/15 refractor supplied to Ross of London (shown below), or the elegant astronomical instruments of Negretti & Zambra (active 1850–1999), which often employed high quality Cooke objectives (Fig. 2.3).

It is difficult to assess how well T. Cooke & Sons penetrated the European market. Interestingly, a study of the origin of the wealthy Hungarian astronomer Miklós Konkoly-Thege's (1842–1916) instrumentation from 1870 to 1910, conducted at his private Budapest Observatory, shows that 66% of the instruments derived from German manufacturers, with only 17% from other foreign countries, mostly England. An interesting trend emerges if one looks more closely at the individual decades. Throughout the 1870s, there were apparently more English than German instruments, due, no doubt, to Konkoly's earlier trip to the English workshops. In the 1880s, the number of German instruments increased, but also the number of instruments made in Hungary. That said, by the turn of the century, German instrumentation dominated.

The decline in British optical ingenuity was well underway by the end of the nineteenth century. But that wasn't the end of the story, for the company was about to be restored to an even greater level of prestige, when Messrs. Cooke took an extraordinary young man under their wing. His name was Harold Dennis Taylor (1862–1943), and his ingenuity became the brain and glory of T. Cooke & Sons in the post-Victorian era.

Beginning his professional career as a trainee architect, Taylor soon became bored and disillusioned with it. He was offered an apprenticeship – which he enthusiastically accepted – with the Cookes at the Buckingham Works. McConnell describes the culture of his new work setting upon the young man's arrival:

> At the time of his arrival, optical design was, as it had been since the time of Thomas Cooke senior, a matter of trial and error based on experience and practice, with only a token nod to theoretical formulae.

The elder Cooke, like all other opticians of his time, probably relied heavily on visual inspection of images through his objectives in the assessment of optical quality. Dr. Jackson describes the 'litmus test for achromaticity' as was then employed by telescope makers:

> The examination of a bright object on a dark background, as a card by daylight, or Jupiter by night, which high magnification powers affords as is well known, the severest test of the perfect achromaticity of a telescope, by the production of green and purple borders about their borders in the contrary case.

Judging by their many happy customers, the Cookes must have done extremely well in their task. That said, Taylor was a different kettle of fish to the elder Cooke. He quickly absorbed the work on diffraction set out by G. B. Airy and established new and higher standards of optical testing and evaluation. Within a year, he had designed a novel kind of photographic exposure meter. Several other patents followed – mainly camera lenses – some of which he sold to Messrs. Cooke outright, and others he received a royalty from.

By 1893, aged just 30, Taylor was placed in charge of all optical projects, followed 2 years later by a seat on the Board of Directors. For the next two decades, Taylor dedicated himself to the advancement of optical knowledge. In 1891, he published a new treatise on refractor optics, *The Adjustment and Testing of Telescope Objectives*, followed in 1906 by a *System of Applied Optics*, which still serve as invaluable resources today. Arguably Taylor's crowning technical achievement was the design and construction of a new kind of refractor objective – an instrument that could be used both photographically as well as visually. Enter the remarkable photo-visual triplet.

By the end of the nineteenth century, the overwhelming majority of public observatories were equipped with large equatorial refractors. The larger instruments, of course, produced a noticeable color fringe around bright objects. Experienced astronomers just learned to ignore it, but the secondary spectrum proved disastrous in long exposure photographic applications. Taylor's new triplet, first produced in 1892, consisted of an outer light baryta flint lens, a middle borosilicate flint element and a light silicate crown comprising the innermost element. An air space was placed between the second and third element. Designed to be used in an F/18 format, the lens produced a wonderfully flat, aberration-free image with color correction an order of magnitude lower than anything seen before. Needless to say, these telescopes proved hugely popular as the new bulwarks of astrophotography, finding their way into observatories on every continent.

Like all dynastic businesses, the end came slowly and unpredictably for Cooke & Sons. In the twentieth century, the firm amalgamated with Troughton & Simms (London) in 1922 to become Cooke, Troughton & Simms. By 1915, however, Vickers had acquired a 70% stake in the business and by 1924, it became a wholly owned subsidiary of the same company.

In the aftermath of the Second World War, Vickers continued to thrive, selling microscopes, surveying equipment and a variety of high precision scientific instruments. Finally, in 1989, the business was purchased by the California-based company Bio-Rad Micromeasurements. Vickers decided to deposit the firm's archives and collection of scientific instruments with the University of York. The instruments are now on display in the Department of Physics, and the archives are cared for by the Borthwick Institute for Archives. The collection also includes a number of printed books, which embody a special collection in York University Library (Fig. 2.4).

Voices from the Grave

Is it possible to divine information regarding the general optical quality of the Cooke refractors that found their way into the private observatories and homes of Victorian gentlemen scattered across the world? One way forward is to explore the comments of historical observers who had used Cooke refractors during the course of their careers.

We shall begin with William Rutter Dawes (1799–1868), revered among double star observers for bringing us his empirical (though as yet unsurpassed by any pseudo-theory) formula used to work out the minimum aperture needed to resolve double stars of a given angular separation. What is less well known is that the reverend was also a first rate planetary observer, apparently possessing extraordinary visual acuity (despite his extreme myopia) at the eyepiece. And he had an interesting purchasing history, having used refractors crafted by Dollond, Merz & Mahler,

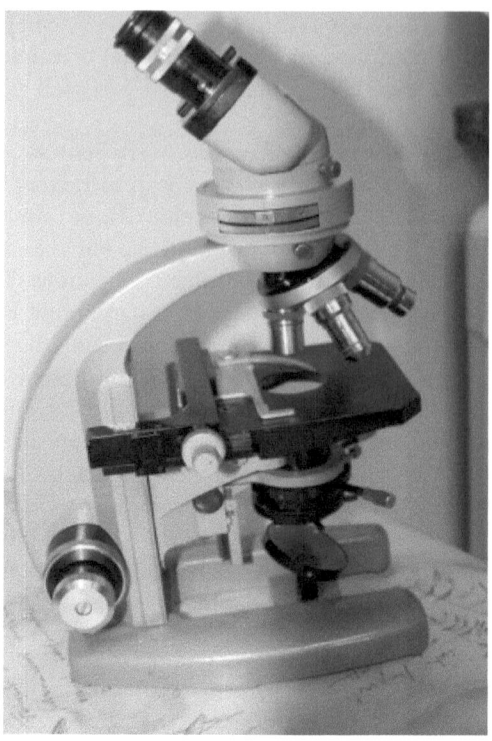

Fig. 2.4 The author's 1960s Vickers binocular microscope

Cooke and even the shining light of American optics, a portrait painter turned telescope maker, Alvan Clark.

Dawes took an interest in Clark's meteoric rise from early on in his career.
Naturally, being an unknown in the industry, Clark at first found it hard to sell his
instruments. What he needed was someone with great astronomical *gravitas* to
champion his cause. If the astronomers didn't come to his telescopes, then he'd
have to bring his telescopes to the astronomers.

And so it was in 1851 Clark wrote to Dawes, describing to him the close double
stars he had observed with his newly-crafted 7.5-in. refractor. Impressed, the reverend sent Clark a more extensive list of close binary stars for him to split, together
with an order for the same objective!

Yet, in the autumn of his life, old 'Eagle Eyes' returned to a Cooke refractor.
Dawes had already made some drawings of Mars in 1862 and at earlier opposi-
tions. In 1864, he used an 8-in. Cooke (that later became known as the
Thorrowgood), usually with a magnifying power of 258×. His drawings, wrote
Richard Anthony Proctor, "are far better than any others…. The views by Beer and
Mädler are good, as are some of Secchi's (though they appear badly drawn).
Nasmyth's and Phillips', De La Rue's two views are also admirable; and Lockyer

has given a better set of views than any of the others. But there is an amount of detail in Mr. Dawes' views which renders them superior to any yet taken." Camille Flammarion concurred: "The drawings by … Dawes brought a new precision to studies of Mars."

And across the Irish Sea, to a beautiful, windswept rural estate near Milltown, County Galway, John Birmingham (1814–1884) used a 4.5-in. Cooke refractor to embark on a special study of red stars, in which he set out to undertake a revision and extension of the best resource of its day on such objects, *Schjellerup's Catalogue of Red Stars*. In all, he included 658 such objects. This work was presented to the Royal Irish Academy in 1876, and its merit was acknowledged by the award of the Cunningham Medal. In 1881 Birmingham discovered a deep red star in Cygnus, which is named after him. He published articles on the transit of Venus and sunspot morphology made with the same telescope, corresponding regularly with the leading astronomers of his day. A lunar crater is named in his honor, too.

Moving next to the Far East of the Empire, at Bankura, India, Chandrasekhar Venkata Raman (subsequently knighted), the recipient of the 1930 Nobel Prize in Physics for his contributions to optical science, was fond of using a 53 Cooke refractor. There is one curious account Raman made while using this telescope to observe Saturn:

> [N]ot only was the Crepe ring an easy object," he says, "but for nearly one hour while the definition was perfect, I made out Encke's marking in the A ring and held it steadily for practically the whole period.

Now, the Crepe ring is quite a difficult target for a 6-in. scope, and for many, 8″ seems to the smallest aperture they'd be happy with. The Enke division (marking) is typically regarded today as a good target for a 10-in. instrument. So, was it fine optics Raman had in his 5-in. Cooke or exceptional eyes – or both? Maybe we'll never know for sure! (Fig. 2.5).

We return, once again, to England, and to the fondly remembered British actor and comedian Will Hay (1888–1949). Though playing the consummate idiot on stage, behind the scenes, Hay was a gentleman of encyclopedic knowledge, with a predilection for astronomical adventure. He set up a fine 6-in. Cooke refractor in a private observatory established at his home in Norbury, London, to study the planets. On the fateful night of August 3 1933, Hay used this instrument and an eyepiece delivering a power of 175× to detect a prominent white spot on Saturn. The spot, located in the planet's equatorial zone, remained prominent for a few days before mysteriously fading away. And although similar phenomena were recorded by earlier observers (Asaph Hall in 1877 and E.E. Barnard in 1903), Hay is credited with the official discovery. Curiously, Hay's beloved 6-in. Cooke, like the spot he discovered, inexplicably disappeared after his death, and, despite diligent attempts to locate it, we are still none the wiser concerning its current whereabouts! (Fig. 2.6).

Hay wrote a wonderful, non-technical book for the newly minted amateur astronomer, *Through My Telescope*, in which his great charm and insight still shines through. A timeless classic if ever there was one! (Fig. 2.7).

Fig. 2.5 The fully restored 8-in. f/16 Fry telescope, at Mill Hill Observatory, London

Modern Perceptions

I have spoken elsewhere of experiences with a couple of Cooke refractors, particularly the 10-in. at Mills Observatory, Dundee, Scotland, which I have peeped through on many occasions during my time in graduate school, and a superlative 4-in. F/18 Cooke-Taylor photovisual instrument. The sharp, contrasty views they both served up were very impressive. But was this representative of what others have experienced? How did these refractors of old settle with folk who have had the pleasure of using them over years and decades? First, Douglas Daniels was contacted, president of the Hampstead Scientific Society, England, who has had the immense good fortune of using the observatory's 6-in. F/15 Cooke since 1967.

White Spot on SATURN, Aug 9th 1933.

W.Hay

Fig. 2.6 The second drawing of Saturn by Will Hay showing the great white spot 6 days after his discovery sketch (Image credit: Martin Mobberley)

Doug spoke about his background and how he became acquainted with Cooke refractors (Figs. 2.8, 2.9, and 2.10).

I have always been a keen lunar and planetary observer and telescope maker since I first became seduced by astronomy at the age of 13 in 1953. I joined the BAA in 1956, which was the year of a very close opposition of Mars. At that time, I had built a 6-inch Newtonian reflector using a mirror made by the late Henry Wildey. I was quite impressed by the performance of this instrument, both on Mars and Jupiter, but I was soon to meet another young BAA member – Terry Pearce. Terry and I became good friends (and still are!). Terry had managed to borrow a 4.5-inch Cooke from the BAA that he had set up in his garden at Chingford in Essex. I was amazed at the sheer size of it. It was an unusual Cooke, two part cast iron column and the equatorial mount was massive for an instrument of that size. But I was even more amazed when I looked through it. The detail on both Mars and Jupiter was astounding – far more contrast than with my 6-inch reflector. That was my first taste of a Cooke.

How and when had Doug first became acquainted with the Hampstead 6-in. Cooke?

In 1967, I joined the Hampstead Scientific Society and was able to use the 6-inch Cooke at the Hampstead Observatory. Again, 1967 was a year with a good opposition of Mars, and the detail observed with the Cooke was so good that I began to attempt photography. I built a special planetary camera with a flip mirror system to keep the planet under close surveillance, waiting for clear moments to make exposures – it was a sort of single lens reflex job but without the lens! (N.B. This was 1967!) My photographs came to the attention of an American student Ron Wells, who was doing a PhD on Martian topography at University College London. Ron was working at the University of London Observatory at Mill Hill – just 15 minutes from my home. I was introduced to the director, Professor Allen, and was allowed to use the 18-inch Grubb – I had the key to the big dome for 6 months. On the same site, there were two smaller domes. One contained the Fry Telescope - an 8-inch Cooke.

Fig. 2.7 A portable 3-in. f/15 Cooke refractor c. 1900 (Image credit: Richard Day)

Once again the Cooke was the instrument that impressed most. On most nights of average seeing, it could easily outperform the 18-inch Grubb. Only when the seeing was excellent could the Grubb show slightly more detail.

Doug was more than happy to recount the telescope's long history.

The Cooke was once owned by a member – George Avenell, and we know that it was in use at the observatory in 1923. It was finally presented to the Society in 1928. Prior to this we have no information. The optical tube appears to have been manufactured around 1900, but we have no hard evidence for this date. When I began using it in 1967, it was mounted on an old Cooke equatorial from a 4.5-inch instrument that was too small. It had the old Cooke falling weight drive and a worm sector – not a complete wheel that was always getting jammed. In the end we built our own heavy duty mount in 1976, driven by a stepper

Fig. 2.8 Image credit: Doug Daniels

Fig. 2.9 Image credit: Doug Daniels

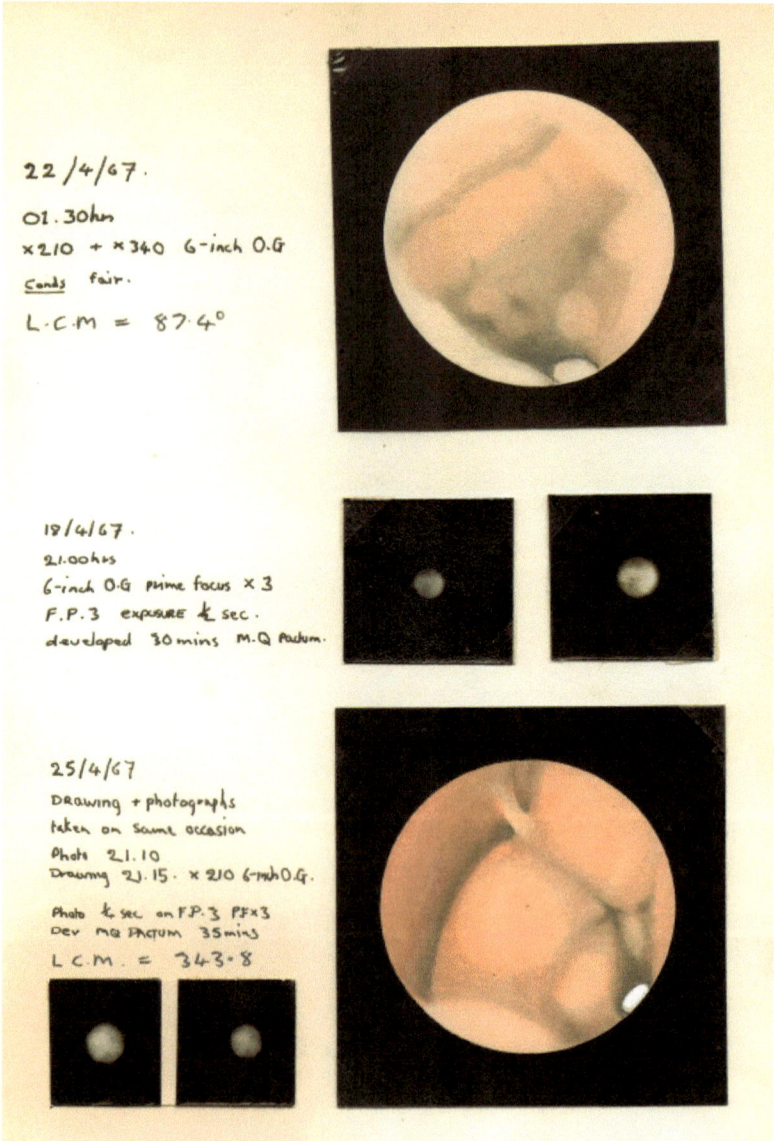

Fig. 2.10 Some of Doug Daniel's recorded details of the Martian opposition of 1967 (Image credit: Doug Daniels)

motor. A couple of years ago, I was in correspondence with Martin Mobberley, who was researching the 6-inch Cooke once owned by Will Hay. I was able to confirm that the Hampstead Cooke was not Hay's instrument.

What about the telescope's maintenance? Is it, in any sense, fastidious in its requirements?

No, not at all. The objective is best left well alone. It gets an annual wipe over with meths and a lint free cloth and every few years is checked for squaring on, which hardly needs any adjustment for long time periods other than that I discourage anyone from touching it. That's another nice aspect of refractors, they are virtually maintenance free, unlike reflectors, which are constantly going out of square and need re-coating every few years.

Doug is no stranger either to the current proliferation of telescope types, each having their advantages and disadvantages. He was asked how he thought the old Cooke faired in the scheme of things:

As a long standing lunar and planetary observer, I have, over the last half century, been able to compare the performance of many different instruments. Given average seeing conditions, I have found that the images produced by the 6-inch Cooke f/15 refractor at the Hampstead Observatory will surpass most if not all other instruments of equal aperture. It will outperform Schmidt-Cassegrains and Maksutov-Cassegrains of 8-inches aperture, and on occasions, it has provided better images of Mars obtained with my own 16.5-inch Dall-Kirkham Cassegrain. It is only on the rare apparitions of excellent seeing that large aperture reflectors can outperform it. I put this down to the absence of a central obstruction that reduces image contrast in all compact reflectors. The current popularity of short focal ratio apochromats is no doubt due to their portability and the need to travel to dark observing sites. But they take longer to acquire thermal equilibrium and require expensive highly corrected short focal length eyepieces to produce sufficiently high magnifications.

Next, we sought the opinion of Dr. Richard McKim, director of the Mars section of the BAA, who has used some of the Cooke refractors in his extensive studies of the Red Planet over the last few decades:

I have used many refractors on a regular basis since the 1970s, 4 cm, 7 cm, 7.5 cm, 15 cm (Cooke, my own), 20 cm (Cooke, Cambridge), 30 cm (Cooke apochromat, Cambridge) and 83 cm (Meudon Observatory). The problem is, I have no basis of comparison with other makes. Until 1988, the Northumberland telescope at Cambridge had a 30 cm Cooke apochromat, as the old lens from the c. 19th had worn so much it was too small to safely fit in the original cell. After the devitrification of one component, Jim Hysom made a new lens for that year, 1988. Equally sharp in definition to my eye, but of course an unfair comparison with an apochromat. Both gave marvelous, sharp images. All I can say is that Cooke achromats and apochromats give marvelous results.

Privately owned Cooke Photovisual refractors, as you might expect, are as rare as hens' teeth. Having looked through a 4-in. f/18 sample, this author can tell you the images of Mars it served up in a recent opposition were nothing short of breathtaking, easiest the finest view of this small planet I have seen through any telescope. Colin Shepherd, an amateur astronomer based in Jervis Bay, Australia, is the lucky owner of a 5-in. F/17 Cooke Photo-visual (c. 1902) but also enjoys observing through his modern ultra-premium 5-in. Astro Physics refractor. So how did he reconcile the old with the new? (Fig. 2.11).

My recollection of its performance is that it delivers images on a par with my 5-inch Astro Physics Starfire (AP130-EFS). I discussed it with my friend, Steven Lee, of the Anglo Australian Observatory, and we both agreed that the performance was similar other than maybe a slight light loss, which is probably due to the lack of coatings on the Cooke PV objective. I have an adapter to permit use of a 2-inch diagonal in place of the original prism, so I can use the Cooke with modern eyepieces. The biggest difficulty with using the 'scope other than its weight is the long tube.

Fig. 2.11 A pristine Cooke lens that saw first light over 150 years ago (Image courtesy of Richard Day)

These telescopes, of course, have long been considered choice instruments for measuring double stars. Curious to see what a contemporary double star observer thought of a big Cooke, Bob Argyle, based at the Institute of Astronomy, in Cambridge, England, is a highly skilled binary star astronomer and author of an influential book on the subject. He has used the 8-in. Thorrowgood (also at Cambridge) refractor for measuring the orbital elements of hundreds of pairs (Fig. 2.12). He said:

> *I'm happy to confirm that this lens is a good as you can expect for the aperture. It was specified for double star work by its original owner, Dawes, and needless to say it fulfills the Dawes limit admirably, separating pairs as close as 0".55 or possibly a little less. On the best nights here, which would not necessarily be regarded as such elsewhere, the disks are perfectly round. In 2004, when Gamma Virginis passed through apastron at 0".37, I was still able to measure the position angle of the elongated image. I gather that J. C. Adams once tried to acquire the telescope from Dawes, and he was of the opinion that it was of better optical quality than the 9.6-inch Dorpat refractor of Struve.*

Animum Debes Mutare, Non Caelum!

What are we to make of all these opinions – both contemporary and historical? For one thing, these are not the words of egregious rogues with hidden agendas! The Cooke refractors clearly delivered and continue to deliver quality views. But, in this age of the Roddier test and interferometer, how well do the 'fine' achromatic Victorian Cooke lenses really stack up? Alas, hard data is not available to answer this question. Prizing one away for laboratory analysis would be rather like trying to acquire a piece of the Constitution! That said, there has been a tendency in

Fig. 2.12 The 8-in. Thorrowgood refractor at Cambridge (Image courtesy of Martin Mobberley)

contemporary amateur culture to assess optical quality using the Strehl ratio inter-pretation, based on laboratory bench tests.

Based on the sample of testaments, many of the larger Cooke refractor object glasses would most probably not be figured to an accuracy much beyond a smooth ¼ wave level at the eyepiece, corresponding to a Strehl ratio not much greater than 0.8 (their peak Strehl at green wavelengths being higher). Contrast that to contemporary, top-of-the-range apochromatic triplets, which can exhibit Strehl ratios higher than 0.95. Yet, as we have seen, many seasoned observers are, and continue to be, more than satisfied with the views these classical refractors of yore deliver. How can these significantly different quantitative assessments of optical quality be reconciled?

One explanation is that we are looking at this question far too simplistically. Doubtless, it is not how well the scope performs in laboratory tests that is at issue here. What counts is how well those quality images are *attained* and *maintained* under typical conditions *in the field*. In previous correspondences with optics authority Vladimir Sacek (www.telescope-optics.net), this author is reminded that, at least in terms of perceived image quality, under field conditions, there's not much difference between a true 1/8 wave and 1/4 wave p-v level optic. Indeed, all the optical greats – Rayleigh, Conrady and Marechal, to name but a few – concluded much the same. What is paramount, however, is the added error, from a number of other sources, and as has been described elsewhere, many of these sources are close to home and may have more to do with the design of the telescope and its thermal management than has previously been acknowledged.

Es Reid, a highly experienced optical engineer based in Cambridge, England, gave his take on this matter:

> *I reckon that, although glass prescriptions and anti-reflection coatings have improved over the years, the methods of final polishing and figuring have not changed to any great extent. A long, heavily mounted refractor should outperform a short one on a lighter, modern mounting. Eyepieces can always be simpler for the same power, and glass will behave thermally as glass always has. Eyes plus brain can select the best wavelength to some extent to give highest acuity in colored images, so I think all of these factors let the old instruments compare very well indeed with modern equipment. It is also interesting that glass companies nowadays have to knock out toxic chemicals, for safety reasons, that can restrict color correction. The inherent and continual smoothness of a refractor wave-front and the high entrance both help a lot in keeping an image sweet!*

Sweet indeed! It seems that quality views delivered by these antiquated telescopes are the result of many things coming together in one package – their unobstructed optics, their simplicity of design, a doublet objective made from thin lenses that acclimate rapidly and completely, their greater elevation off the ground, well away from any sources of ground turbulence and anthropogenic heat and their great depth of focus, making accurate focusing easier and, of course, their generous image scale (1/f). Couple that to the ruthless genius of the human eye to 'filter' the signal from the noise, introduced by the purplish unfocused light of the standard achromatic prescription, and you can easily understand why they would delight a patient or experienced observer.

Such attributes, of course, make them ideal as measuring instruments and that's precisely why, it is likely, they were built with such enthusiasm. Indeed, J.B. Sidgwick, writing in his influential book from 1971, reminds us yet once again of the advantages of these long focus refractors. They enjoy, he says, "greater independence of temperature variations, with steadier images and higher possible magnifications than with a reflector of the same aperture," (pp. 420–21). Seen in this light, these telescopes are not the 'extinct dinosaurs' contemporary astronomical culture would have us believe. As we have seen, there is every indication that these instruments are 'highly adapted specialists', supremely useful in the noble art of astronomical mensuration.

In retrospect, it seems daft (it really does!) that these instruments should ever fall out of favor with amateur astronomers. But that, sadly, is the reality on the ground. Doug Daniels thinks he knows why.

At the end of the day, I think the reason that the long focal length refractor has 'fallen out of favor' is simply due to the 'long focal length.' It was (is) an instrument ideally suited for a permanent observatory and to be used for visual observation. Both of these applications seem to be out of favor today. Because of the proliferation of light pollution, users want portable equipment to drag to dark locations, some going as far afield as Bermuda or the Canary Isles. Because of the comparative ease afforded by digital imaging, users demand faster focal ratios. Because of the 'competition' amongst enthusiasts to produce ever more astounding images of 'deep sky' subjects and the relatively small size of CCD chips, users demand wider fields of view. None of these requirements can be met with an f/15 refractor. BUT, if you are interested in observing and drawing planetary detail, or observing and measuring double stars, and you have the space to build a 10-foot diameter dome in an unobstructed pollution free location, then the 'old fashioned' achromatic refractor is hard to beat. If my astronomical 'fairy godmother' could wave her wand and grant my wish, it would be for an 8-inch f/15 Cooke refractor made at the end of the nineteenth century at Bishop Hill in Yorkshire.

Although the Sun has long since set on both the Roman and British Empires, the legacy of Thomas Cooke lives on in the people who have had the pleasure of using these fine telescopes, whether a gentleman's 3-in. glass or a large observatory class instrument. Small wonder that Cooke's erudite obituarist was compelled to write in 1868, "[To] our English Fraunhofer.... Whose science and skill had restored to England the pre-eminent position she held a century ago in the time of Dollond (Fig. 2.13)."

The ancient Greek philosopher, Socrates, once remarked that all learning consists of that which is pre-existing in memory. Having recounted the allegory of Thomas Cooke & Sons to you and extolled the virtues of some of their optical wares, we hope that this will all seem familiar to you, too!

Fig. 2.13 *In Amatam Memoriam*: The author's fine 4-in. f/15 refractor, inspired by the workshops of T. Cooke & Sons (Image by the author)

Chapter 3

Once upon a Time in America

We have loved the stars too fondly to be fearful of the night.
John A. Brashear

In comparison to Great Britain and Germany, progress in telescope making in the New World was painfully slow. Indeed, as we have seen, the largest refractor in the United States before 1830 was a 5-in. Dollond achromat. The paucity of public observatories across the nation in the early nineteenth century is evidence enough that the country had not yet fully exploited her penchant for astronomical adventure. America needed a great lens maker, and it found its answer in a Massachusetts portrait painter named Alvan Clark.

Like John Dollond and Thomas Cooke, Alvan Clark also came from humble origins. Born in 1804 in Ashfield, Massachusetts, he was the fifth of ten children of Abram, himself a descendant of whalers from Cape Cod, and Mary Basset Clark. After receiving his formal education at a small grammar school located on the family farm, he was set to work with his older brother making wagons. Shortly afterwards, he discovered his latent talent for art. Indeed, by the last year of his teens, Alvan had grown proficient at engraving and drawing. By 1824, he had produced an impressive portfolio of work that he carried with him to Boston, where he eked out a meager living, traveling through the picturesque Connecticut Valley, creating portraits in ink and water color.

It was in these formative days of youth that Clark became exposed to the occupation that would secure his immortality as one of the finest telescope makers the world has ever bore witness to. He made the acquaintance of a one Edward Hitchcock, an evangelical pastor who was also rather evangelical about his unworldly passion; amateur astronomy. By the age of 21, Alvan realized he needed a steadier income and signed up as a professional engraver with Mason & Baldwin of Boston, subcontractors to the Merrimac Manufacturing Company. In his spare time, he continued to paint portraits but also took to studying astronomy and attended lectures presented by Warren Colburn, a superintendent of the Merrimac Company.

It was here also that he met the love of his life, Maria Pease, whom he married in March 1826. They enjoyed over 60 long years of life together. And she bore him four children – two daughters, Maria Louisa and Caroline Amelia, and two sons, George Bassett and Alvan Graham.

Clark continued in his engraving job until 1836, when he decided the time was right to return to portraiture. Accordingly, he moved his family to Cambridgeport, a district of Boston, and opened up a small studio from which he quickly established himself as a first rate artist, receiving commissions from high profile academics and politicians. He even painted a portrait from a daguerreotype of a one W. R. Dawes – a man who would elevate him to worldwide fame.

We do not know the precise circumstances under which Clark took to the task of telescope making. His first projects apparently involved reflectors with apertures up to 8 in. in aperture. Indeed, during the winter of 1847–1848, Clark decided to test the optical quality of a freshly polished 7.5-in. speculum by making detailed drawings of the Orion Nebula (M42) while refraining from consulting any previous drawings. His sketches came to the attention of William Cranch Bond, director of Harvard College Observatory, who not only lauded their quality but also noticed that Clark had in fact plotted stars that were previously unknown and, indeed, had even escaped the attention of the darling of visual astronomy, Sir William Herschel, who carried out similar surveys using his 20-ft telescope.

However, Clark soon lost interest in constructing Newtonians on account of the poor reflectivity of the speculum metal used at the time, and in 1847, he began to figure his first lenses from discarded objectives prized from old or abandoned telescopes. As anyone who has performed such a task knows, it's a very time consuming activity. But his patience paid off. Unlike Cooke and Fraunhofer, Clark's approach to practical optics was more intuitive than theoretical. That much became clear when he was first granted an opportunity to look through the great 15-in. Harvard refractor in 1848. It was a moment that was to change the course of his life. In his memoirs, Clark wrote:

> I was far enough advanced in the knowledge of the matter [optics] to perceive and locate the errors of figure in their 15-inch glass at first sight. Yet, these errors were very small, just enough to leave me in full possession of all the hope and courage needed to give me a start, especially when informed that this object glass alone cost $12,000.

And start he did, closing his art studio to master the art of figuring old lenses. But Rome wasn't built in a day, as he discovered to his dismay, when he sent some of his early works in glass to the Bonds at Harvard College. One 4-in. instrument of Clark's design apparently yielded comatic images that greatly sullied his early reputation as an optical craftsman and hence his relationship with Bostonian academicians for several years to come. Nor did Clark's disposition endear him easily to others. He was neither academically distinguished nor endowed with great business acumen. Indeed one source claimed that "Mr. Clark's lack of mathematical learning, or learning of any kind, kept him out of the confidence of the scholarly persons around Boston."

Contrast that to his contemporary and fellow telescope maker, Henry Fitz, who had established a thriving business supplying achromatic objectives for newly established observatories that were springing up and down the country.

Born in Newburyport, Massachusetts, in 1808, when Henry was 11 years old, his family moved to Albany, New York, and later to New York City. Henry's father was in the printing business, and accordingly, the young man dutifully served his apprenticeship in the family workshops. Fitz had a penchant for practical mechanics and enjoyed tinkering with the machinery in the shop. When he was 19, he decided to leave the family business and take up a new line of work as a locksmith. Fitz was a keen amateur astronomer and read widely in the field. In his spare time he began to make and build mirrors and lenses to house in his homemade tubes, and in his early thirties, Fitz transformed his hobby into a thriving business, specializing in the manufacture of small refracting telescopes.

As his reputation grew, Fitz began to receive commissions to build larger, more ambitious telescopes for both well to do private individuals as well as leading American universities and institutions. Indeed, between 1840 and 1855, it has been estimated that Fitz made about 40% of all telescopes sold in the United States! Among his larger telescopes were two 13-in. instruments, one commissioned by Allegheny Observatory in 1861, which is still in use today and the other for the Dudley Observatory in Albany, New York. The latter instrument apparently went into private hands around the turn of the century but fell into obscurity thereafter.

Fitz also apprenticed a number of future, independent telescope makers like John Byrne, who flourished in New York during the late 1800s, furnishing a number of small equatorial refractors for the discerning American amateur. One of Byrne's instruments, a fine 5-in. f/15 glass, was acquired in 1877 by the prodigious young Edward Emerson Barnard. Though he bought it at the heavily discounted price of $330, it still represented a sum of money equivalent to two thirds of his annual salary at the time. But with his 'pet,' as he so affectionately referred to it, Barnard discovered his first comet in 1881 (one of many it transpired) and embarked a lifelong study of the planets.

Henry Fitz was a socialite of the first order. Clark, on the other hand, apparently suffered from an inferiority complex. Indeed, one gets the distinct impression that he sometimes doubted his own optical abilities. Why else would a man not know how to correctly price his instruments or not advance his own cause by showing off his wares at the various international exhibitions?

Alvan Clark's first 'serious' instrument had a 5.25-in. aperture, followed by an 8-in., both of which were as good as any of European origin. Naturally, being an unknown, he at first found it hard to sell his instruments. What he needed was someone with great astronomical gravitas to champion his cause. If the astronomers didn't come to his telescopes then he'd have to bring his telescopes to the astronomers! (Fig. 3.1)

In 1851, Clark wrote to the prominent English amateur astronomer, the Reverend William Rutter Dawes (1799–1868), describing to him the close double stars, most

Fig. 3.1   Alvan Clark, Sr. (1804–87)

notable of which was 95 Ceti, which he had observed with his 7.5-in. refractor. Impressed, Dawes sent Clark a more extensive list of close binary stars for him to split, together with an order for the same object glass!

With his Clark refractor, Dawes later wrote that he had enjoyed the finest views of Saturn he had ever seen. Indeed, Dawes purchased no less than five Clark refractors in his career. Clark's reputation in England spread like wildfire, and he soon received another order, this time for a 8-in. object glass from a certain William Huggins, who had used the lens, mounted on a massive equatorial platform designed by T. Cooke & Sons, as the centerpiece for his pioneering work in astronomical spectroscopy conducted between 1860 and 1869.

Over the coming years, Clark and Dawes corresponded frequently, becoming good friends in the process. In 1869, Clark paid a visit to his English friend, attending a gathering at the Observatories at Greenwich and a meeting of the Royal Astronomical Society, where he was introduced to Lord Rosse (of Leviathan fame) and Sir John Herschel. These meetings did much to cement Clark's reputation as an instrument maker of the highest order (Fig. 3.2).

To this day, very little is known regarding Clark's methods for producing his lenses. Like the Dollonds of the previous century, they left no records of their procedures. But nothing was done in secret, either. The factory often welcomed curious visitors. One snooty caller opined that the methods employed were crude and inferior to those used by European standards. But Alvan Clark never professed himself to be an optical theorist. He had, as we have seen, a highly developed intuition for

Fig. 3.2 A nicely restored 9-in. Clark refractor made under the aegis of Carl Lundin in 1915 (Image credit: Siegfried Jachmann)

crafting some of the best refractors in the world. He could apparently detect tiny irregularities on the surface of the lens and often retouched it using his bare thumbs while examining the image at the eyepiece. We do know that polarized light was often used by many nineteenth century telescope makers – the Clarks included – to inspect their optical glass and the finished lens. The test was as simple as it was telling. Inhomogeneous glass would usually reveal streaks or splotches, whereas a well-made blank would not (Fig. 3.3).

As well as these preliminary tests, the Clarks used traditional methods to assess the optical figure of their lenses. Alvan Graham, in particular, was a well-respected observer and used his instruments to resolve difficult double stars. At the beginning of Clark's telescope making career, obtaining large, high quality glass blanks was difficult. But necessity is the mother of invention, and Clark took to refiguring old and ill-figured lenses to fill out his order books. In a delicious irony, the archives

Fig. 3.3 The majestic 8-in. Clark refractor at the Chabot Space & Science Center, Oakland, California

show that quite a few of those refigured object glasses originated in the workshops of a one Henry Fitz!

In terms of design, most of the Clark objectives are similar, though not identical, to the Fraunhoferian blueprint, consisting of an equiconvex crown (R1=R2) and a meniscus flint in which R3 is made a few percent shorter in radius than R2. R4 (closer to the eyepiece) becomes a long-radius, convex surface being almost flat. Thus, the Clark is an air-spaced design, similar to the Fraunhofer, but with weaker curves. R1, R2, and R3 are all close in radius to one another. Spherical aberration can be cancelled (corrected), just as in the Fraunhofer design. In addition, if R1 and R2 become reversed during cleaning, there is no apparent change in performance.

As Deborah Jean Warner notes in her book, *Alvan Clark & Sons, Artists in Optics,* there was nothing especially distinguished about the Clarks' methods of working their object glasses:

> *The blank disk was first polished so that it could be tested for purity and eveness – a piece of glass too heavily striated would be rejected. The grinding and preliminary polishing were done on a rudimentary machine which consisted simply of a horizontal turntable rotated by steam power. The table held the tool, a cast iron lap the same curvature as the lens, but reversed. The lens was held in the rotating lap and slowly moved about. Small lenses could be worked by one man, a larger lens was fitted with four wooden handles by which two workmen, walking around the table, could give the proper movement. Alternately, the lens was supported from a horizontal beam which mechanically imparted a reciprocating motion. The early Clark lenses were ground with emery, but by 1887 the Clarks were using cast iron sand as an abrasive because it had a lesser tendency to break down. When the rough grinding was finished the metal lap was exchanged for one of pitch and the lens was polished.*

The Clarks also employed other state-of-the-art techniques to fine tune the performance of their objectives. Calling on the new methods of 'local correction' described by French physicist Jean Bernard Leon Foucault in 1859 and some practical pointers from Henry Fitz, the Clarks were known to re-touch all four surfaces of the object glasses. Indeed, the Clarks were the first to introduce the double pass auto-collimation tests to their optics (Fig. 3.4). As Warner also notes:

> *The Clarks later developed and used a test twice as sensitive as the original one. Light from a point source was focused by the lens and then reflected by an optically plane mirror back through the lens to an eyepiece close to the light source. Since light passed twice through the lens, the effect of any irregularities was doubled.*

As news spread of the incredible discoveries the Clark telescopes were making in the hands of astronomical evangelists, it wasn't long before the commissions for Clark telescopes came flooding in. His first major order was a 18.5-in. instrument for the University of the Mississippi. In a rare splash of self confidence, Clark sold his home to invest in new premises – at Cambridge, Massachusetts – to build and test object glasses. Accompanied by his two sons, he constructed a 230-ft-long tunnel to 'bench test' the optical prowess of his objectives on artificial stars.

On a freezing late January night in 1862, the Clarks were performing routine tests on a newly completed 18.5-in. object glass. They were trying to gauge how much off-axis glare the instrument exhibited by timing how long the light from Sirius was perceptible before the star was in view. Although the bright star was still behind a corner, Alvan Graham noticed an eighth magnitude 'spark' appear a full 3 s before Sirius came into view. This was the first-ever recorded observation of Sirius 'pup' and a testament to the quality of optics used to divine its presence.

The Clarks went on to build the largest and finest refractors the world had ever seen, which include the 24-in. refractor at Lowell Observatory, used to divine the Martian canals in the colored imagination of Percivall Lowell (1855–1916), the 26-in. instrument at the U.S. Naval Observatory used by Asaph Hall to discover the asteroid moons of Mars and which is still in use today by professional astronomers as a dedicated double star instrument, and the 36-in. Lick refractor,

Fig. 3.4 A 1919 vintage Clark refractor (Image credit: Dan Schechter)

used productively by legendary observers such as Edward Emerson Barnard and Sheldon Wesley Burnham in California. Things were seen with the latter telescope that scarcely anyone has witnessed since. Indeed, it was with the great Lick refractor that both the eagle-eyed Barnard and John. E. Mellish were said to have glimpsed craters on the Martian surface. Saturn's tiny, elusive moon, Amalthea, was also discovered by Barnard with the Lick (Fig. 3.5).

The largest Clark refractor of all, the 40-in. at Yerkes Observatory, Williams Bay, Wisconsin, might well have been larger, but it is doubtful that it was ever as good as the 36-in.. Barnard apparently disagreed though, claiming that there would always be the occasional evening where the larger aperture of the Yerkes refractor

Fig. 3.5 The 40-in. Clark refractor at Yerkes Observatory, Wisconsin (Image credit: University of Chicago)

would show more than its sibling perched high on a mountain in the Golden State. The enormous weight and extreme difficulty in casting, figuring and polishing such large lenses meant that refractors had reached their natural limit in terms of size. Reflectors would go on to win that prize. For the record, a 49-in. lens with a focal length of 187 ft was also made by the Clarks, but subsequent tests revealed it to be rather poor optically.

In recognition of his work, the elder Clark received four honorary Master of Arts degrees and medals. Arguably, the most illustrious accolade bestowed upon him came in 1874 when Harvard College Observatory made Clark a Master, conferring on him the title, *Artificem egregium, speculatorem rerum, coelestium callidum.*

Part of the success of the company can be attributed to the sheer longevity of Alvan Clark (Sr.). He took an active part in the business right up until a few years before his death in 1897, aged 83. George Bassett passed away in 1891 followed by his brother Alvan Graham in 1897. With no suitable blood heirs, a new firm, Alvan Clark & Sons was incorporated in 1901. The business was taken over by a talented Swedish optician and mechanician, Carl Axel Robert Lundin, who became chief instrument maker for the Clarks in 1874. Lundin died in 1915, and Sprague-Hathaway Manufacturing Co. acquired all assets and staff of the Alvan

Clark & Sons Corporation in 1933. Finally, in 1958, the demise of Sprague-Hathaway represented but the final act in the dissolution of the company, and Alvan Clark & Sons was no more.

Alvan Clark & Sons produced many smaller telescopes for the 'gentleman astronomer.' Probably the most famous of these is the 4-in. f/15 instrument used by the celebrated astronomer-author William Tyler Olcott. Made in 1893, it had a wooden tripod that supported the brass and nickel tube and a hand driven worm wheel. Olcott later bequeathed the instrument to Phoebe Haas, who, in turn, passed it onto the celebrated Walter Scott Houston, the late *Sky & Telescope* columnist of "Deep Sky Wonders." Sadly, the instrument has seen little in the way of starlight ever since.

Another celebrated amateur who put his faith in a Clark refractor was Sheldon W. Burnham (1838–1921), who quickly established himself as one of the most prolific double star discoverers of all time. Burnham began his astronomical career with a modest 3.75-in. refractor, but found both the mount and the aperture didn't quite meet his needs. But after a chance meeting with Alvan Clark in Chicago in 1869, he ordered up a 6-in. equatorial refractor of his highest quality.

"I told them what I wanted, and what I wanted it for," Burnham wrote:

Every detail was left entirely to their judgment, stipulating only that its definition should be as perfect as they could make it, and that it should do on double stars all that it was possible for any instrument of that aperture to do.

In due course of time this instrument was delivered, and was set up in an observatory prepared for it in the meantime. My attention for some reason or other, which I am unable to explain, had been almost exclusively directed to double stars previous to this while using the smaller telescope referred to. This preference was not in any sense a matter of judgment as to the most desirable or profitable department of astronomical work, or the result of any special deliberation upon the subject. It came about naturally, without any effort or direction upon my part.

Daniel Schechter, a physician based in California, is an avid collector of antique Clark refractors, including a wooden-tube 4-in. dating from 1860 and a later instrument of like aperture manufactured around 1919 (Figs. 3.6, 3.7, 3.8, and 3.9). When asked if he could discern any differences in optical quality between them he said:

The 1919 Clark is a 10 out of 10 and the 1860 wooden tube Clark is a 9.5 out of 10. I had the wooden tube Clark objective expertly collimated, spaced and tested. It tested better than 1/8 wave after expert spacing and collimation. The problem with the wooden tube Clark is the wooden tube. The faceplate that the focuser screws into was badly bent when I obtained it. A friend of mine took most of the bend out and with shims I can almost, but not quite center the focuser on the front objective. A Cheshire eyepiece shows the two circles in contact and about 90% aligned. Since there are no collimating screws, I have to be happy with almost aligned. The 1919 Clark is perfectly aligned. Just like the Brashear, I think the 1960 objective would be a close equal to the 1919 Clark objective if I could perfectly center the focuser and align the objective. Please note that I am nitpicking. Most anyone would be very happy with the 4 1/8-inch wooden tube Clark.

The optical and mechanical quality of the Clark refractors guaranteed their success in the United States and abroad. But how good were the Clark lenses? Indeed, more broadly, the curious individual may legitimately inquire as to the differences

Fig. 3.6 A 4.13-in. Clark (c. 1860). Note the wooden tube (Image credit: Dan Schechter)

between the quality of nineteenth century glass and their modern counterparts. As we have seen, Clark reflected skylight – which is partially polarized – off the back side of his lenses, so that it traveled through the glass twice, and then viewed it with a Nichol prism to reveal striae and other imperfections that would not show up otherwise.

On examination of a few lenses originating from the nineteenth century with a 10x loupe, it is fairly common to see tiny bubbles in the glass that are all but absent in contemporary refractor objectives. By and large, their presence is not unduly damaging to the overall image but simply reflect limitations of the technology available to early glass workers. Owner testimonies, while informative, can only tell half the story, though. What is required is an objective testing procedure that can quantify how good these objectives really were. To that end, Dick Parker was consulted, who has ground and built his own telescopes for decades and now runs

Fig. 3.7 Check out the objective of this 4-in. Clark (Image credit: Dan Schechter)

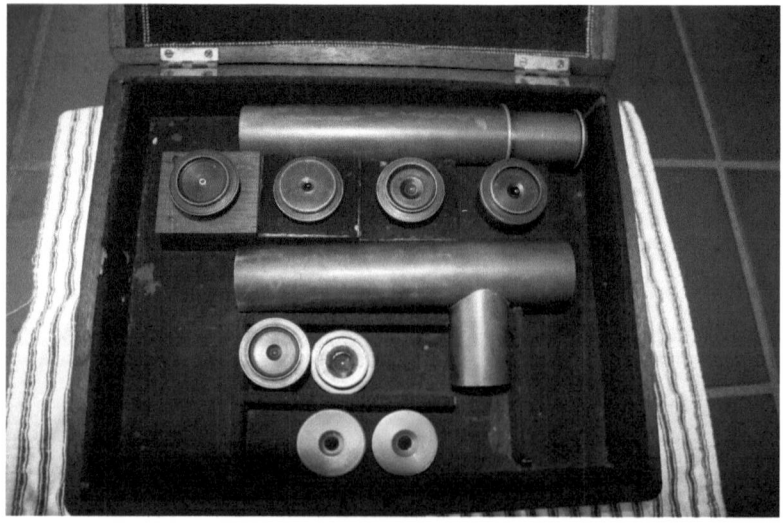

Fig. 3.8 Check out this box of goodies that came with Dan's 1860 Clark (Image credit: Dan Schechter)

a telescope-making workshop at his U. S. home in Tolland, Connecticut (Fig. 3.10).

Dick recounted details of work he carried out on the testing of a 5 in. f/15 Clark objective:

Fig. 3.9 The 4-in. Clark secure in its original wooden case (Image credit: Dan Schechter)

Fig. 3.10 The 5-in. (127 mm) lens, made by the Alvan Clark & Sons Corp., being tested

In September, 2010, I had an opportunity to test a 5-inch diameter achromat lens, which is currently owned by an antique telescope collector. It is dated to c.1915 and appears not to have ever been installed in a telescope.

Dick made use of the so-called an auto-collimation test, which yields a null return for an optic bringing light originating at infinity to a single focus point.

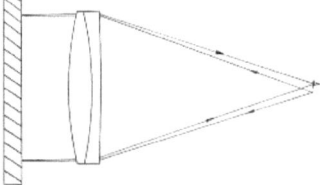

Fig. 3.11 Schematic diagram of the auto-collimation test. The auto-collimation flat is to the left. The light source and viewing by eye are to the right with the eye slightly above, as shown

Collimated light, in optical terms, means light rays that are parallel. The object glass makes its own collimated beam to simulate light coming from stars at infinite distance. Curiously, this test was actually adopted by Alvan Clark & Sons, and they reported this testing technique in *Scientific American* Supplement No. 932 for November 11, 1893. Dick took some time to explain the procedure (Fig. 3.11).

> *For a practical description of the test, a pinhole source of light (artificial star) is placed at the focus of the lens. The light travels to the lens in a diverging cone, then passes through it, where it is refracted to a parallel bundle of rays (collimated). The collimated light then is intercepted by an optically flat mirror and reflected back through the lens, where it is refocused back near the original source. At this focus location, the returning focused beam can be examined by an eyepiece, knife edge, Ronchi screen or other suitable testing device.*

"I chose to use a Ronchi screen," he said, "which consists of a series of consecutive opaque and transparent lines, etched in at a very fine spatial frequency. I adopted a screen with 133 lines per inch, which was placed just behind (further from the lens) where the returning beam comes to focus and examined. Then the screen was placed just before the beam comes to focus and re-examined. If the rays from the lens are brought to a perfect focus, what will be seen is a series of lines across the lens that will be straight, parallel, and equally spaced. Any curvature of the lines will reveal a defect in the lens. The views with the screen just behind the focus (further from the lens than the focus) should be the same as views with the screen in front of where the rays come to focus. The light source used had a wavelength 565 nm (green)."

Figures 3.12 and 3.13 show the test return with the Ronchi screen just behind and just in front of where the rays come to focus, respectively. What should be visible by comparing the two is slight bowing of the lines. Notice that the lines appear to bow inwards toward the outer part of the lens in Fig. 3.12 and outward toward the center of the lens in Fig. 3.13.

As is shown here, light coming through the center of the lens has a shorter focus than light coming from the outer parts of the lens. This is called spherical aberration.

What is also noticeable in Fig. 3.12 is increased exaggeration of this bowing toward the center third of the lens. This is an indication of a local condition where the focus at the center of the lens is just a bit shorter than it would be from spherical aberration alone. Figure 3.13 was made with the Ronchi screen at focus so that one bar of the screen acts as a single knife edge. This makes for a very sensitive test for

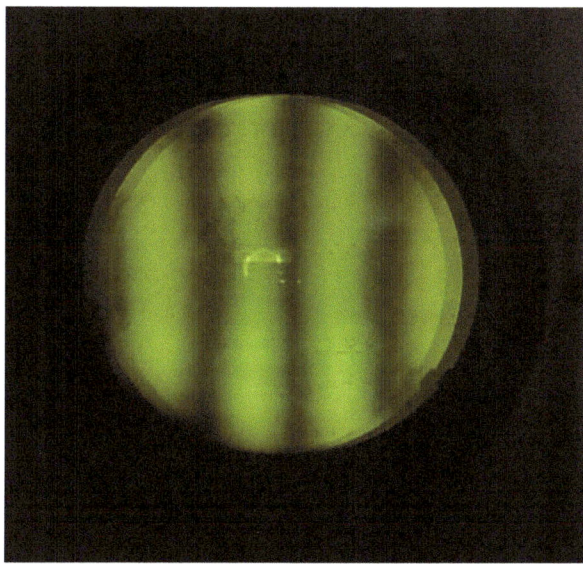

Fig. 3.12 Ronchi screen behind focus showing outward bowing toward the outer part of the lens

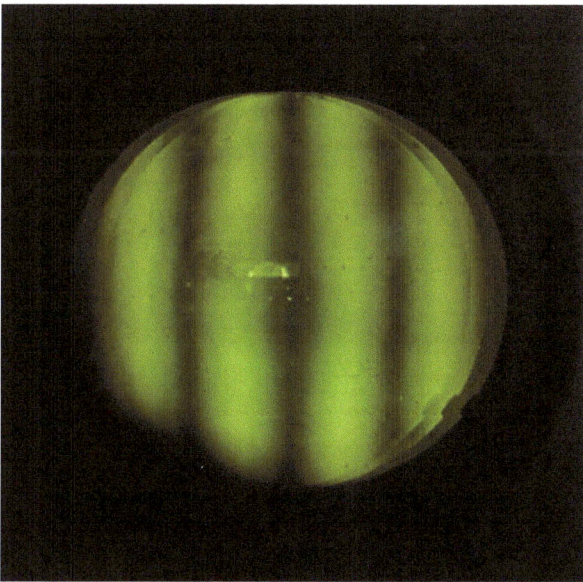

Fig. 3.13 Ronchi screen in front of focus showing outward bowing toward the center of the lens

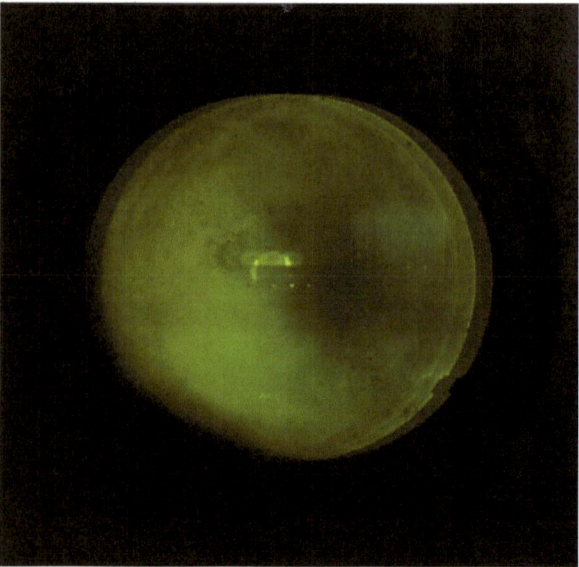

Fig. 3.14 Ronchi screen at focus so that one bar acts as single knife edge, revealing the short focusing center

local conditions. If all the rays were focusing at a common point, the surface would appear uniformly gray. Notice the center appears as if there were a depression in it. This is indication of a short focusing center (Fig. 3.14).

What does this tell us about the lens? It means that the lens is not "perfect," but it should provide very fine images when installed in a telescope. By comparing to a known optic with an optical path difference between the center and edge due to spherical aberration equal to a quarter of the wavelength of light, tested the same way by auto-collimation, this lens would have a spherical aberration just about a quarter of a wave or better.

So, according to Dick Parker's optical bench tests, this particular Clark lens met or exceeded the minimum quality contemporary opticians would deem useful. That's quite an acceptable figure, given that the company was probably working to the minimum standards arrived at by Lord Rayleigh in 1879. Indeed A. E. Conrady's later book, *Applied Optics and Optical Design,* which was published in 1929, arrives at much the same conclusion.

The Instruments of John Brashear

The astounding success of the Clark telescope-making saga doubtless cast a long, inspirational shadow over new telescope makers springing up all across the United States. Prominent among the *homines novi* was the great Pennsylvanian optician

John Brashear (1840–1920), who designed and built excellent instruments ranging in size from 4-in. equatorial refractors for the amateur astronomer to 30-in. reflective behemoths used by the astronomical professionals.

Born in Brownsville, Pennsylvania, a small town 35 miles south of the sprawling city of Pittsburgh, John's father, Basil Brown Brashear, was a saddler by trade, and his mother, Julia Smith Brashear, was a dedicated school teacher. As a lad, John Brashear had the immense good fortune to come under the aegis of his maternal grandfather, Nathanial Smith, who repaired clocks for a living but, as luck would have it, was also an enthusiastic amateur astronomer, who took him to observe the sky through some good telescopes owned by the 'Squire' Joseph P. Wampler, who set up his traveling telescopes in Brownsville. Those virginal views of Luna and the Ringed Planet, Saturn, had a profound influence on the young Brashear, who developed a strong interest in astronomy.

After receiving a common school education until age 15, he took up an apprenticeship to a machinist and by age 20 had fully mastered the trade. The following year, Brashear moved to Pittsburgh and spent the next 20 years there working as a millwright. In 1861, he met Phoebe Stewart, a Sunday school teacher. They fell in love, married in 1862 and remained together for the rest of their lives. Of too little means to purchase a telescope, Brashear took to building his own in a small coal shed turned workshop located in his back garden.

By 1870 Brashear had completed his first telescope – a small refractor – and immediately opened his doors to neighbors, friends and strangers to observe the sky through it. Dr. Samuel Pierpont Langley, the director of the Allegheny Observatory, encouraged him to establish a proper workshop for astronomical instruments. The business expanded, and by 1880 he founded the John A. Brashear Co. with his son-in-law and partner, James Brown McDowell. Brashear's business went from strength to strength, quickly establishing itself as an internationally renowned maker of fine optics.

Brashear was the first of the great nineteenth-century opticians to meticulously record his work for others to follow. His name is associated with many innovations, yet remarkably, he never sought to patent any of them. For example, he introduced a novel way of silvering mirrors (discussed in a later chapter), known appropriately enough as the Brashear process – which was not surpassed until vacuum metalizing replaced it in 1932.

In 1890s, Brashear made his second of three trips abroad, visiting the great optical houses and observatories of Europe and conducting a lecture tour. In 1898 he became director of the Allegheny Observatory in Pittsburgh, continuing in this post until 1900. From 1901 to 1904, he was acting chancellor of the Western University of Pennsylvania, now known as the University of Pittsburgh, having served as a member of the board of trustees since 1896. Brashear was also a trustee of the Carnegie Institute of Technology and served as president of the Academy of Science and Art.

In 1919, he suffered a debilitating bout of food poisoning, which left him gravely ill for 6 months. He died the following year, at age 79, while at his South Side home. He was survived by a daughter and several siblings (Fig. 3.15).

Fig. 3.15 The 20-in. Brashear refractor at the Chabot Space and Science Center, Oakland, California

Brashear made many large, observatory-class telescopes – reflectors as well as refractors – during his career. Arguably, one of his sweetest is the 20-in. Warner-Swasey refractor. Amateur astronomer Jared Wilson, from Piedmont, California, gives up some of his free time as an attendant at the Chabot Space and Science Center at Oakland and was kind enough to provide some interesting facts about the 20-in. Brashear refractor, known more affectionately known as 'Rachel.' "She was built for Chabot in 1915 or so, and displayed at the World's Fair that year in San Francisco," he said, "then moved over to Oakland for her permanent installation. Warner and Swasey were the prime contractors to mount the great Brashear tube. Rachel's most famous moment was tracking the *Apollo 13* spacecraft in order to provide data to calculate the final burn for re-entry on that failed mission. My understanding is that she was used because she was one of the few large telescopes left with a mechanical clock drive whose tracking rate could be modified enough to

Fig. 3.16 A 4-in. f/15 Brashear achromat atop a stylish but sturdy alt-azimuth mount (Image credit: Dan Schechter)

keep up with the proper motion of the spacecraft. None of the then-current electric drives could handle anything other than sidereal, lunar, or solar tracking."

As well as his great observatory behemoths, Brashear sustained a thriving business selling small instruments for an adoring amateur market. And Brashear refractors are alive and well in the twenty-first century. Daniel Schecter, who we met earlier in connection with the Alvan Clark collectibles, is also a fan of Brashear refractors. He is the lucky owner of a c. 1900 4-in. f/15 Brashear. Schechter was asked how his 4-in. f/15 Brashear achromat stacked up against his Clark refractor of similar specification. "Many people who own both Clarks and Brashears claim that the Brashear is slightly better," he said, "especially in terms of color correction. I have mounted the 4-in. f/15 1919 Whitman Clark and the 4-in. f/15 Brashear on a side by side saddle a few times. Early on they seemed equal, but lately I would give the nod to the Clark. I am considering sending the Brashear objective to an expert to get it expertly collimated and spaced. My guess is that after I do that, they both will give 10 out of 10 star tests. I can't imagine the Brashear outperforming my 1919 Clark, though, because that objective is as close to perfect as I and many others have viewed through (Figs. 3.16 and 3.17)."

Many (but not all) the refractors built by Brashear, like those fashioned by the German astronomer and telescope maker Carl August Von Steinheil (1801–70), had the flint element placed ahead of the crown. Some claim that this configuration yielded slightly better results than the standard crown-first configuration. The precise reasons for Brashear's departure from the conventional are quite unknown. One guess is that the type of flint used in Brashear's refractors was more weather resistant than its counterparts today (Fig. 3.18).

Fig. 3.17 A 'one-armed astronomer' with Brashear refractor (Image credit: Dan Schechter)

Fig. 3.18 Brashear often employed a Steinheil-like approach to building his object glasses. This 4-in. Brashear achromat has a flint leading element

Mogey, Melish and Tinsely

The late nineteenth century saw the introduction of a number of other players in the American telescope-making industry. The German-born American brothers Lohmann established a business in Greenville, Ohio, where they produced a number of finely crafted astronomical refractors – mostly equatorial refractors – very similar in form and function to those produced by Alvan Clark & Sons. They were eventually taken over by another telescope maker, Warner and Swasey. Also highly prized among collectors are the instruments of William Mogey, who in 1882 turned his practical knowledge of optics into a business opening a workshop at 418 West 27th street, New York.

William first produced camera lenses, but as demand for his works grew, his younger brother David joined the company and expanded their product lines to include surveyor's transits, spectroscopes and refracting telescopes.

In 1893, W. & D. Mogey moved to larger premises in Bayonne, New Jersey. There the company remained until 1911, when the bad light pollution and city smog forced the company to move to the darker, cleaner skies of Plainfield, New Jersey. By the 1920s the new firm, William Mogey & Sons, had established a world-class reputation for producing fine telescope optics for a global market. Curiously, despite the fact that reflectors were easier and cheaper to manufacture, Mogey senior apparently had no real commercial interest in producing reflectors. Of refractors he wrote, "They are to be preferred because of their compactness, portability and efficiency."

Typically, his instruments came in an f/14 format, employing high quality achromatic lenses ranging in aperture from 3 to 12 in.. These were sold either as bare optical tube assemblies or fully assembled on large equatorial mounts. This author has recently seen a 4-in. equatorial Mogey fetch $5,000 at auction (Fig. 3.19).

Another sought-after name in classic American telescopes is Tinsley. Founded in 1926, the firm was based in Berkeley, California, and specialized in the manufacture of fine refractors, and Newtonian and Cassegrain reflectors, ranging in size from small amateur telescopes to large observatory-class instruments. They also sold telescope-making kits (reflectors and refractors) using top-quality components for the practical-minded amateur (Fig. 3.20).

Tinsley was one of the earliest companies to develop astrographs for serious amateurs in the early 1950s. Its 8-in. Cassegrain, for example, came complete with a 3-in. refractor guide scope (shown below) and provided a rock steady platform to take celestial shots using the then popular 2.25 in. × 2.25 in. reflex cameras. And, if you thought the base price of $995 was too much, Tinsley offered a 3-year payment plan to boost sales.

Other notable, though less celebrated, telescope makers emerged in the United States during the first half of the twentieth century. Of particular interest is the Wisconsin born John Edward Mellish (1886–1970), who, in his day, was a highly respected telescope maker and comet discoverer. Like so many of the other great telescope makers so far discussed, Mellish's story is one of the ordinary man made

Fig. 3.19 A classic Mogey 4-in. refractor on a beautiful mahogany mount (Image credit: John Miles)

great. Growing up on his maternal grandfather's farm, he completed but 4 years of elementary education before being put to work on the land. This pastoral lifestyle was ideally suited to the quiet and introverted young man. The dark, pristine night skies of rural Wisconsin Mellish had access to throughout his youth endeared him to astronomical adventure.

When he was 16 years old, Mellish acquired his first spyglass for the princely sum of $4. This was followed soon after by a 2-in. refractor of higher quality with which he enjoyed countless hours of stargazing. But Mellish soon yearned for larger instruments that would unveil more and more of the universe.

Mellish wanted to build his own telescopes. After teaching himself for a while, he decided to embark on his first telescope-making project. He ordered two glass disks some 6 in. in diameter from a firm in Chicago and spent the winter grinding a mirror. The result was a quality 6-in. Newtonian reflector with which he discovered his first comet in 1907. The discovery reached the ears of Edward Emerson Barnard,

Fig. 3.20 A 1947 vintage 4-in. f/15 Tinsley refractor (Image credit: Clint Whittmann)

the amateur astronomer turned professional, whose keen eyes had secured him a salaried position at Yerkes Observatory. Indeed, Barnard invited Mellish to Yerkes to acquaint himself with its talented staff.

Encouraged by his spectacular discovery, Mellish pursued more ambitious telescope projects that supplemented his modest income from farm laboring. In his early twenties, Mellish's optical skills were so refined that he began to publish articles in prestigious journals such as *Popular Mechanics* and *Scientific American*. The publicity he received from those publications were apparently enough to secure ongoing business from large observatories and universities.

By 1915, having discovered no less than three comets, Mellish accepted a position as an observer at Yerkes, receiving a small grant – arranged through Edward C. Pickering at Harvard College – to cover his living expenses and the replacement of a farmhand.

In his spare time, however, Mellish resumed the business of making telescopes to order, and commissions for his optical wares soon came flooding in. In 1928, he finished one of his largest orders, a fine 10-in. refractor that remained in use at Sherzer Observatory, at the University of Michigan, until 1989, when a fire razed the observatory and the instrument with it. Mellish also built large Cassegarin reflectors. One of his finest, a 29.5-in., was commissioned by the University of Illinois.

Despite his success as a telescope maker, Mellish maintained a keen interest in active observing throughout his life. As we have noted earlier, the acclaimed visual astronomer E. E. Barnard had reported seeing Martian craters between 1892 and 1893 with the 40-in. Clark refractor at Yerkes. Using the same telescope, Mellish also reported seeing them in November 1915.

Some details of those observations emerged in a letter to a one Mr. A. W. Leight, Philadelphia:

> There is something wonderful about Mars, it is not flat but has many craters and cracks. I saw a lot of the craters and mountains one morning with the 40 inch and could hardly believe my eyes and that was after sun rise and Mars was high in a splendid sky and I used a power of 750 and after seeing all the wonders I went to Barnard and showed him my drawings and told him what I had seen and I had never heard of any such thing having been seen, and he laughed and told me he would show me his drawings made at Lick in 1892–93 and he showed me the most wonderful drawings that were ever made of Mars, the mountain ranges and peaks and craters and other things both dark and light that no one knows what they were. I was thunder struck and asked him why he had never published these and he said no one would believe him and would only make fun of it. Lowell's oases are crater pits with water in them, and there are hundreds of brilliant mountains shining in the sun light. Barnard took whole nights to draw Mars and would study an interesting section from early in the evening when it was just coming on the disk until morning when it was leaving and he made the drawings four or five inches diameter and it is a shame that those were not published.
>
> I do not know as anyone would be allowed to even look at them now, they are at Yerkes and will stay burried (sic) I suppose. The canals are not as straight as I drew them but the best I could do because the fine detail was just at the limit of vision and waves wash across (sic) and destroy detail and then it comes at instants. I have done a lot of work making mirrors and lenses for Lowell Obs. but when it comes to Mars they do not see anything in my drawings that they ever saw on any map, at least that showing a wonderful lot of detail. Well to say the best for it I could hardly make out anything. So that is the way it goes. For planetary detail one must have a very long focus mirror or objective, a short focus will not give the needed contrast.....

In retrospect, it seems that Mellish's association with this remarkable 'discovery' adds quite a bit to the collectability of his telescopes. After all, who wouldn't like to own a telescope made by a man whose keen eyesight showed him features on the Red Planet that scarcely another soul could see? (Fig. 3.21)

American telescope makers, of course, would go on to build the largest instruments on Earth and take astronomy to new heights of sophistication, including celebrated names such as G. W. Ritchey (1864–1945) and James Gilbert Baker (1914–2005), to name but two. Indeed it would be the United States that both raised and set the bar for astronomical research for much of the twentieth century (Fig. 3.22).

In the next chapter, we return to the heart of Europe and explore the extraordinary instruments made by Carl Zeiss and why they are rightly adored by amateur astronomers and telescope collectors alike.

Fig. 3.21 John Mellish pictured beside one of his fine refractors

Fig. 3.22 A 'riot' of American-made refractors set up for the 1878 solar eclipse at Forth Worth, Texas (Image credit: Dan Schechter)

Chapter 4

Zeiss Is Nice

The nineteenth century witnessed the flourishing of great optics houses, both in Britain and in the New World. As we have seen, the extraordinary success of Thomas Cooke in England and Alvan Clark in the United States owed much of their success to technology that was first formulated by German opticians, particularly the pioneering work of Joseph Von Fraunhofer. Were it not for his untimely death in 1826, Fraunhofer might well have gone on to become the greatest name in telescope optics ever. Alas, that was not to be. That being said, Germany's talent for producing fine optical wares was restored through the extraordinary accomplishments of Carl Zeiss and the optical company he founded.

Like his father before him, Carl Friedrich Zeiss (1816–1888) began his career as a toymaker. But this occupation proved to be of little lasting interest to the bright young man, who had acquired an interest in practical optics. He served an apprenticeship in the shop of Dr. Friedrich Körner, which honed his skills with fine tools and precision machinery used to fashion microscopes and other scientific instruments. In the winter of 1846, Zeiss decided to set up a small workshop at Neugasse 7, Jena on the Saale River in the district of Thuringia in Germany. It was a cunning business plan, for he quickly secured contracts to repair many optical instruments from the local university. In addition to that, in the first year of operation, Zeiss managed to sell some two dozen microscopes, which gave him encouragement and the confidence to expand his business.

Within a year, Zeiss moved to a larger premises at Wagnergasse 32 and also hired his first apprentice. Innovations came swiftly from his workshop. In 1857 the company introduced their first compound microscope, known to collectors as the "Stand I." By the early 1860s Zeiss compound microscopes were considered to be "among the most excellent instruments made in Germany," having been awarded the Gold Medal at the Thuringian Industrial Exhibition for his optical nuances.

N. English, *Classic Telescopes: A Guide to Collecting, Restoring, and Using Telescopes of Yesteryear*, Patrick Moore's Practical Astronomy Series, DOI 10.1007/978-1-4614-4424-4_4, © Springer Science+Business Media New York 2013

By 1864, the company had expanded to such an extent that a still larger premises was required, and, accordingly, the Zeiss plant moved to Johannisplatz 10, employing a workforce of about 200 people. In 1866, the company had delivered its 1,000th microscope. But Carl Zeiss didn't rest on his laurels. He realized that to stay ahead of the game he had to oversee still greater improvements in the optical quality of his instruments and that meant the total abandonment of the inefficient trial and error processes that had traditionally been employed by his optical forebears. He couldn't do it alone, however, and, to that end, he found latent talent in a young man called Ernst Abbe (1840–1905).

Abbe grew up in abject poverty, his father having worked as much as 16 h a day to support his family. Abbe's intellectual brilliance, however, earned him his way through school by gaining scholarships as well as with some assistance from his father's employer. As an undergraduate, Abbe studied physics and mathematics at the University of Jena and was later awarded a doctorate in thermodynamics from the University of Göttingen.

In 1863 Abbe joined the faculty at the University of Jena and was introduced to Carl Zeiss in 1866. Abbe became very interested in the optical challenges facing microscopy. Late in 1886, Zeiss and Abbe formed a partnership whereby Abbe became the director of research of the Zeiss Optical Works. Abbe laid out the framework of what would become a cornerstone of modern computational optics. Among Abbe's most significant breakthroughs was the formulation in 1872 of a wave theory of microscopic imaging that became known as the "Abbe Sine Condition." This approach made possible the development of a new range of 17 microscope objectives designed via mathematical modeling. In Abbe's words:

> [B]ased on a precise study of the materials used, the designs concerned are specified by computation to the last detail – every curvature, every thickness, every aperture of a lens – so that any trial and error approach is excluded.

As mentioned above, before the collaboration between Zeiss and Abbe, lenses were made by trial and error. However, these objectives were the first lenses ever made that had been designed entirely on the basis of advanced optical principles. The comparatively high performance of the new Zeiss microscope objectives earned international acclaim for the company.

Working first with the microscope, Abbe realized that he needed to find improved glass types if he was going to make progress in correcting the chromatic aberration found in the achromatic doublet objectives. In January 1881 Dr. Abbe met with the chemist and glass technologist Friedrich Otto Schott (1851–1935), and together they pursued a scientific approach to the determination of raw ingredients to be used in glass formulations and developing manufacturing techniques that would produce hundreds of new types of both optical and industrial glasses.

Their collaborative work would also see the improvement of molten glass mixing and the annealing processes. In 1882 Schott moved to a new glass-making laboratory set up for him in Jena, and in 1884, he formed the Schott and Associates Glass Technology Laboratory. From there, he developed many new glass types, a number of which are still in use, including borosilicate crown, known to us today as the indispensible BK series of glasses. Schott's glass innovation made possible the introduction by Abbe in 1886 of the first "apochromat" lens, and by 1890 he had succeeded in filing a successful patent for an apochromatic triplet objective.

As we have seen previously in connection to Thomas Cooke & Sons, Harold Dennis Taylor traveled to Jena in 1895 to visit and learn from Ernst Abbe. At the Schott glass works, he attempted to obtain a regular source of supply for some of their special optical glasses. These new glasses would cancel out secondary spectrum that had hampered attempts at photography using the standard achromatic refractors. Abbe's pioneering work on apochromats produced doublets for use in telescopes, but the glass prescriptions he used were quite unstable and therefore unlikely to survive years of exposure to the elements. Taylor's approach was different, and the so-called photo-visual triplets he designed employed glass that was more durable than Abbe's.

Progress on the apochromatic microscope objective came earlier. In 1886 Zeiss marketed the first objective made from a veritable 'magic substance' known commonly as 'fluorspar,' which is a crystalline form of calcium fluoride (or more simply, fluorite). Abbe discovered that by mating optically clear, polished fluorspar with other, more traditional glasses in a microscope objective, all traces of false color could be removed. So secret was the use of fluorite that Abbe marked an "X" on the data sheet for the fluorite element, so as to hide its remarkable optical properties from the prying eyes of other ambitious opticians. When the academic world first learned of them, the new apochromatic objectives sold like hot cakes, with Zeiss, naturally, absorbing nearly all of the high end market.

Carl Zeiss passed away peacefully on December 3, 1888. His son had entered the business many years before and held the reigns of corporate power for just a year before retiring himself. The business was thereafter incorporated as the Carl-Zeiss-Stiftung in 1889 and continued to go from strength to strength by head hunting the finest minds in European optics. Indeed, Zeiss retained an almost legendary reputation for the manufacture of optical instruments of all kinds, which resonates even in the twenty-first century.

Zeiss is associated with some of the greatest names in optics. If you're an eyepiece fan, you'll probably be familiar with Albert Koenig (1871–1946), father of the highly regarded Koenig ocular. After studying mathematics and physics at the Universities of Jena and Berlin he became acquainted with Ernst Abbe, who served as his Ph.D. supervisor. In October 1894, Koening, the newly minted doctoral graduate, joined the firm. After his arrival at Zeiss Jena, Koenig's talent and industry was immediately appreciated and rewarded with rapid promotion. By the turn of the century, Koenig was a key person behind the development of numerous optical systems, including eyepieces, prisms, and telescopic objectives. Indeed, he dedicated over half a century to the company, pouring his great intellect into an array of new optical gadgets.

Koenig was responsible for the development of several novel eyepiece designs, some of which presented apparent fields of view of up to 90°! There were several types made: combinations of singlet and doublet lenses, and with varying glass prescriptions. And while there are some contemporary makers who advertise a

"Koenig" or the more corrupted "Konig" eyepieces, the design was never really finalized. The Koenig eyepiece has a concave-convex positive doublet mated to a convex, positive singlet. With apparent fields of view from 65° to 70°, they were (and still are) very popular owing to their generous eye relief. Indeed, only the designs of Al Nagler (born 1935) have surpassed them in this regard, but only with the aid of many additional elements. This author has used them to good effect with telescopes of longer focal ratios.

Another famous Zeiss employee was Heinrich Erfle (1884–1923), who in 1921 patented a practical design for a wide-angle ocular. Erfles usually consist of five elements comprising two achromatic doublets, with a convex singlet element sandwiched in between. Erfle conceived of the design during World War I, primarily for military purposes. Erfle eyepieces are designed to have a wide field of view (about 60°) but suffer from astigmatism and ghosting, especially in the shorter focal lengths. That said, Erfles remain quite popular owing to their large eye lenses and good eye relief (Fig. 4.1).

Prior to 1935 all refractive lenses were uncoated. That is, they presented their polished surfaces to the surrounding air. Because some of the light is reflected off the glass surface this cuts down (4–6% per surface) on the percentage of light transmitted to the eye. The more glass elements used, the more light is lost. Seen in this light, one can easily understand why the early eyepiece designs were simple!

The first glimmer of a breakthrough in solving this problem came after something curious was noted in 1886 by H. Dennis Taylor. After carrying out careful tests comparing the light transmission of old, tarnished glass with new, 'clean' objectives, Taylor discovered, to his great surprise, that some of the older, tarnished lenses had the greater light transmission and seemed to reduce ghosting in the images! What's more, the tarnished layer had a refractive index (a measure of how

Fig. 4.2 Zeiss were world masters in binocular manufacture (Image credit: Richard Day)

much light is bent while passing through a transparent material) between that of glass and air. The tarnished layer clearly had the effect of reducing the amount of light lost by reflection off the glass surfaces.

A proper understanding of this phenomenon took a few more decades to unravel, when in 1935 the Ukrainian-born Olexander Smakula, an optician working for Zeiss, learned how to apply very thin coatings of magnesium fluoride (MgF_2) to the surfaces of the lenses, decreasing light loss due to reflections from 4% to just 1%. These so-called *anti-reflection coatings* actually remained a German military secret until the early stages of World War II. Today, they allow multi-element optical devices to be made by minimizing light loss.

By the early twentieth century, Zeiss had successfully negotiated limited partnerships with overseas companies. It was a good strategy in an evolving market that was, ostensibly, increasingly global in outlook. By manufacturing a product in the country wherein it would be sold, it enabled the company to bypass import tariffs, as well as expediting the delivery of products to their customers.

And they made a fortune selling binoculars, too. Indeed, by the beginning of World War I, Zeiss Jena had developed a total of about 59 models of hand-held binoculars for consumer and military use. Pioneering research funded and conducted by Zeiss technicians revealed that a typical adult eye when dark-adapted would dilate to about 7 mm diameter. After considering the efficiency of visual optics in low light applications, the company introduced the first 7×50 mm binocular prototype (one of the most successful models ever) in 1910 (Fig. 4.2).

The Zeiss 'Scopes

Although Zeiss produced all kinds of telescopes for both the amateur astronomer and observatories, it is their small refractors that are most commonly sought after by collectors. One reason for that is the durability of refractor optics compared to reflecting telescopes. Designing large telescope objectives employing fluorite was a technological impossibility when it was first conceived in the late nineteenth century. The know-how to produce large crystals of the mineral and the tools required to polish the soft crystal to standards exhibited in their crown and flint achromats simply hadn't been invented. That said, Ernst Abbe found ways to improve the achromaticity of the refractor by experimenting with other types of glass.

Arguably one of the most notable of his astronomical telescope achievements was the Zeiss "B-Objektiv" (Type B Objective), an f15 air-spaced triplet apochromat incorporating BaLF4/KzF2/K7 glass elements. Made in apertures from 2.4 to 8 in. (60–200 mm), the Type B exhibited great color correction and was devoid of spherical aberration. Indeed, it remained well regarded from the turn of the century until well after World War II.

The principle reasons why it fell out of favor with astronomers were twofold. Firstly, the glass proved especially susceptible to weathering. In addition, the elements had an annoying tendency to become de-centered relative to each other, owing to the strongly curved interior surfaces and the large air gap between the middle short-flint element and the final crown. Even small alignment errors would produce strong coma in the image. To ameliorate the problem, Zeiss set about redesigning more robust lens cells, introducing carefully made spacers and retaining rings that expanded and contracted with temperature in such a way as to keep the lens elements properly oriented with respect to one another. Indeed, one later optician quipped, "One must…expect from the user that he value the B-objective like a highly sensitive physical measuring instrument…and treat it accordingly."

By 1933 Zeiss had manufactured several proven refractors of the "E," "A" and "AS" achromatic doublet designs, and apochromat triplets of the "U.V." and "B" (Koenig) designs. These telescopes were offered in apertures of up to 65 cm (25.6 in.).

An inspection of the Zeiss official catalogs from the 1920s reveals similarities and differences between these objectives. The E objectives, for example, were doublet Fraunhofer designs, typically in a f/15 format and employing silicate glasses. The A objectives were also of doublet design but had elements made of lower dispersion glass for better color correction in f/15 relative apertures (Fig. 4.3).

Fig. 4.3 Zeiss also produced rich field refractors such as this 110 mm f/7 achromat (Image credit: Richard Day)

Highly prized among amateur collectors and active observers alike are the AS series of refractors, which were manufactured in apertures ranging from 2.5 to 8-in. monsters. The label "AS" stands for "*Astro-Spezialobjektiv*." The design of this object glass is credited to the Zeiss optician August Sonnefeld. At its heart is a doublet objective made from a BK7 crown and KzF2 'Kurzflint.' The AS series also differs from the traditional Fraunhofer design in that the flint is leading, that is, it's a Steinheil design. Performance-wise, the AS refractors would now be considered to be semi-apochromatic, and users report that they compare very well to modern ED scopes.

These telescopes were put together with great attention to detail. The lenses, for example, were figured to an accuracy of between 1/10 and 1/12 wave peak to valley across their surfaces. What's more, the lens cells, which were constructed from either brass or aluminum, were compensated for thermal variations and shocks. That said, the Zeiss AS 'scopes were not without their problems. Many owners of particularly old AS models have noticed a tendency for the delicate spacers between lens elements to wear down, introducing slight imperfections into the images (Figs. 4.4 and 4.5).

How do the classic Zeiss AS 'scopes stack up against their modern equivalents? Alexander Kupco, a particle physicist and avid amateur astronomer based in the Czech Republic reported a curious comparison between his newly acquired Zeiss AS80 f/15 and his state-of-the-art short tube Stellarvue SV80s f/6 triplet apochromat.

> One night I put those two telescopes side by side on the same mount. I was curious if I could see any difference between those two excellent 80 mm refractors; an SV80S is an 80/480 mm (f/6) apochromat with a well regarded LOMO triplet. I use this 'scope mostly

Fig. 4.4 The Steinheil doublet objective of the Zeiss AS80 (Image credit: Pat Conlon)

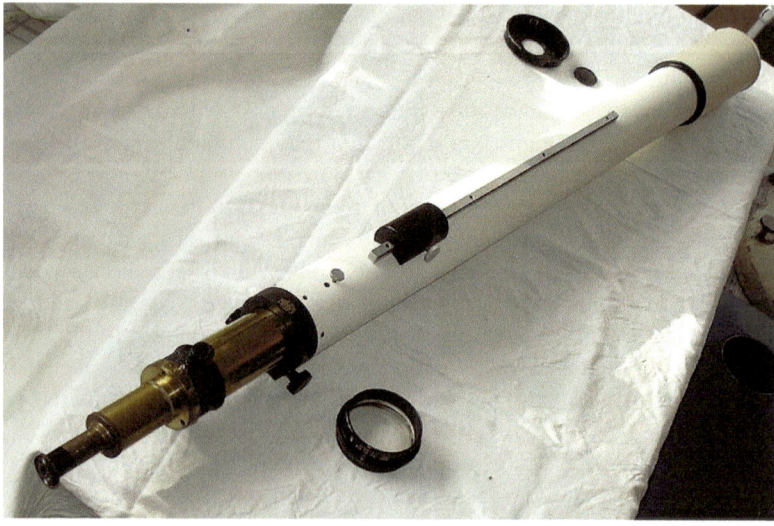

Fig. 4.5 The Zeiss AS80 ready for first light (Image credit: Pat Conlon)

for astro-imaging. The Zeiss AS80/1200 (f/15) is a semi-apochromatic Steinheil doublet with reduced chromatic error with respect to a classical achromat (due to KzF2 + BK7 glass combination). Both telescopes star test very well and with a green filter. I'm not able to trace any significant deviation from textbook patterns. (It does not mean that they aren't there, as I'm just learning about star testing).

The main object of study was the Moon. It was quite low, and seeing was strongly affecting the views already at about 100x magnification. I used Pentax XO5.1 (94x) for SV80S and CZJ ortho O-12.5 (96x) for AS80/1200. When I started, the difference between the telescopes was huge. SV80S came nowhere near the performance of Zeiss. Although the image in Zeiss was boiling in this magnification, minute details were still clear, the image was very sharp, very tonal and 3D-like. The SV80S was a little bit brighter, and it may have had a little more color neutral image, but the details were washed out. At this point, the 'scopes were outside only for about 15 minutes. I store them at home, and the temperature drop was about 10 C. I repeated the observation after one hour and SV80S performance greatly improved, the image getting closer to Zeiss, but still the Zeiss was obviously better. Not that I could find some detail which I could not observe in SV80S, but they were popping out to the eye in Zeiss, and once I knew about them I was able to find them also in SV80S. In general, I had a feeling that I was observing with the SV80S at lower magnification. The visibility of details was similar when I switched from a Zeiss to a CZJ O-16 (56x) eyepiece.

Similar performance could be seen on Delta Cygni. The AS80/1200 was showing the two components clearly for 100% of time at 171x (TMB Mono 7 mm), whereas as in SV80S at 186x (Pentax XO2.5), the second component looked in the beginning only like a brighter part in the first diffraction ring. This was after the telescopes were 30 minutes outside. After another 30 minutes the image in SV80S greatly improved and the double star was an easy catch as well, although the image in Zeiss was still more steady.

I could see similar behavior on Izar. Here I noted slightly bigger color contrast between the two components visible in SV80S. Again Zeiss was showing the double a little bit more clearly and sharply – the arcs of the first diffraction ring could be seen all the time (Pickering 6/10) while I would judge the seeing in SV80S at least one step smaller 5/10 (arcs were seen only occasionally).

In summary, it shows that even this small triplet cools down really slowly, and it takes about one hour with 10 C temperature difference before one can start to do really critical observations at high powers. But even after one hour, the long Zeiss was still better, both on the Moon (this I would call a pretty visible difference) and double stars (here I would call it a tiny difference). Either the triplet was not fully thermalized even after more than one hour or there is indeed some advantage in average seeing conditions for long doublets.

What a bizarre account! What could be the explanation for the apparent advantage of the older, longer focus Zeiss AS80 in comparison to the state-of-the-art apochromatic SV80s triplet? The answer lies in the design of the objective. The triplet has a greater lens mass, with more curved elements than the simpler Zeiss doublet. This makes the latter take considerably longer to cool off. In addition, the Zeiss uses glass types with a considerably lower coefficient of thermal expansion than the low dispersion glass employed in the triplet. That means the Zeiss is more resistant to changes in shape as it cools and so keeps its focus better. Indeed, these attributes, as we shall explore more fully in a later chapter, go quite some way to explaining the remarkable properties of classical refractors over their shorter focus apochromatic counterparts (Fig. 4.6).

Fig. 4.6 The AS 100/1,000 of 1960s vintage, astride a majestic equatorial mount (Image credit: Max Lattanzi)

The School 'Scope

By far the most common Zeiss telescope to fall into the hands of amateurs across the world is the little 63 mm Telementor. Introduced by Zeiss in 1972, the instrument replaced a similar but older pedigree AS 63/840 semi-apochromat that made its debut in the aftermath of World War II (Fig. 4.7).

The newer model possessed more conventional glass (BK7/F2) and was intended to find its way into every school in the Germany, hence its name *Telementor,* which

Fig. 4.7 The perennially popular Zeiss 63 mm Telemator (Image credit: Richard Day)

means 'school telescope.' It is a cemented achromatic (C type) doublet with a focal length of 840 mm (f/13.3).

The Telemator is in fact the same telescope but included as part of an observing package. You get the optical tube, a 7×42 rotatable finder, a nicely machined, motorized equatorial mount on a wooden tripod, an accessory tray and a 1.25 or 2 in. diagonal holder.

In its aperture class, the Telementor is virtually without peer. To its credit, Zeiss maintained very high standards even with these modest 'scopes. It is rather unfortunate that they will slowly be boxed away, and fewer will savor the sweet taste of starlight. They were meant to be used.

Fortunately, one can still pick these instruments up at reasonable prices, either as an optical tube assembly or with their dedicated mounts. Keep an eye out for

Fig. 4.8 The cemented achromatic doublet objective of the Zeiss Telementor (Image credit: Richard Day)

them on the online classified adverts and eBay. Although fakes are rare, they occasionally show up, as we'll explore later in the book. *Caveat emptor!* If possible, try before you buy (Fig. 4.8).

As we have seen, in the years after World War II, Zeiss introduced a growing line of amateur optics, lens cells and complete telescopes, for the amateur market. But that wasn't the end of the story. In the 1990s, Zeiss brought to market their APQ series of refractors, arguably the finest instruments ever made available to amateur astronomers and highly desirable to collectors today.

The APQs were oil spaced fluorite triplets built to the very highest standards and were offered for sale in 105 mm, 130 mm and 150 mm formats. Curiously, a 206 mm f/8 APQ telescope was also advertised for a short while, but Zeiss was never able to complete a single instrument before the small telescope manufacturing group at Jena were instructed to reduce the output of the line. Employing what were then the most advanced scientific testing methods, each optic was rigorously tested before release. Indeed, the late refractor lens designer Tom Back once recounted how he had heard from a long-time Zeiss employee that each APQ was given its final test by a senior optician who took every 'scope out to check its performance on the skies. If the optic did not meet his standards, it would be rejected out of hand, irrespective of its laboratory test report. Zeiss would not take the chance of having a single sub-standard optic with its logo on it leave the factory. Zeiss APQ refractors ceased production in 1994, and they've become collector's items ever since.

One enthusiastic owner of the Zeiss 100 mm f/10 APQ said:

> *It gives stunning high power views of Mars and Jupiter. Jupiter looks like an etching with 8 plus bands and white spots, etc., easily visible. Saturn is amazing with detail on the globe*

and in the rings of extremely delicate contrast. This 'scope reveals Antares companion in poor conditions easily. The 5th and 6th stars in Orion Trapezium are always visible. It also performs well on deep sky way ahead of other 4-inch 'scopes in its class. Double stars closer than 1.13″ can be seen although not separated as kidney bean-like shapes. People literally cause a stampede to look through this lovely telescope at star parties. especially those in the know. The only sad thing is I don't own this modern work of art. The mount it sits on is very easy to use and stable, very stable. The clock drive has a periodic error of around 5-10″, which allows 10 minute photos with no guiding if polar aligned well.

Gary Beal, an amateur astronomer based in New Zealand is an avid Zeiss fan. Here is what he had to say about his own Zeiss acquisitions:

Many memories abound with my relatively short time owning Zeiss 'scopes, but one that sticks in the mind is the 2001 view I had of Mars. It was quite well placed for southern hemisphere observers. I used the Telementor 2, and this was of course on the T mount, and Zeiss tripod. I recall using the O-6 and a Mars glass filter, and the views were just superb. Obviously a good night of seeing as well. I was accompanied by a good friend and equally enthusiastic Zeiss owner Kevin Barker, and I recall him using his Zeiss APQ100 on the Ib mount. The views were better in the APQ, but not greatly, not on this night anyway.

Fig. 4.9 New Zealand observer Gary Beal with his collection of Zeiss refractors (Image credit: Gary Beal)

Successive years meant the Telementor 2 was sold, and I acquired my current two Zeiss scopes, a gorgeous AS63/840 with the small Zeiss GEM residing in the storage case with the optical tube (and accessories), and an equally wonderful ED80/840 in its original case. Both are destined to go to the grave with me. I love them; quirky, wonderful to use, and giving an immense sense of pride of ownership not to mention the history, which makes owning of a Zeiss 'scope so special.

In briefly surveying the long and illustrious history of Zeiss one cannot help but feel a great sense of pride in owning and using an instrument that originated from its workshops. After all, how many companies survived two world wars, the Cold War and the turbulent changes of the post modern world that attended German re-unification and the global financial meltdown? Zeiss can proudly say they've been there, done that (Fig. 4.9).

Chapter 5

The Story of Broadhurst, Clarkson and Fuller

We have previously recounted the extraordinary success of the Dollond and Cooke dynasties and their elevation to the highest echelons of optical prowess throughout the eighteenth and nineteenth centuries. But it would be misleading to think that these were the only success stories to emerge from Britain. Indeed, as we have alluded to earlier in the book, many skilled artisans grew up around the success of these international names. Foremost amongst them is the story of Broadhurst, Clarkson and Fuller, which continued the traditional methods of constructing and repairing handmade telescopes until the dying days of the twentieth century.

The company was founded by the polymath Benjamin Martin (1704–1782), a teacher, lexicographer, and scientific instrument maker of some considerable repute. Martin moved from Surrey to Fleet Street in 1750 mainly to be near the Royal Society so that he could see and hear the latest ruminations of his hero, Sir Isaac Newton, who gave lectures there on a regular basis. In 1749 Martin published *Lingua Britannica Reformata,* which contained a universal dictionary that predated, by a half dozen years, the famous dictionary of Samuel Johnson.

Noted for his innovative designs, Martin was lauded as one of the pioneers of the modern microscope. But he was also a noted spectacle maker, famous for inventing 'Martin's Margins' – a novel style of "visual glasses" with inserts surrounding a small, round lens. These inserts, usually fashioned from horn or tortoiseshell, restricted the amount of light reaching the eyes, protecting them from glare. Martin was also the first to use colored lenses to aid people with reading difficulties. And although ill-appreciated at the time, it has subsequently been employed to aid in the treatment of some forms of dyslexia (Fig. 5.1).

Joined by his son, Joshua Lover Martin in the 1770s, B. Martin & Son was established and began manufacturing and selling a wide range of scientific instruments, including Hadley's quadrants, spectacles, microscopes and telescopes.

Fig. 5.1 Benjamin Martin (1704–1782) (Image credit: Steve Collingwood)

Lacking the acumen to run a successful business, Martin hired managers to run the firm so that he could continue his lectures around London, eventually retiring in 1776. But things came to a head in 1782, when Martin, now at the advanced age of 78, was declared bankrupt due to poor management of the firm. The wretch attempted suicide, dying only a month later from the wound. The business was swiftly auctioned off and bought by Charles Tulley, whom we met earlier in connexion with Dollond.

As we saw earlier, Tulley was a respected optician and instrument maker who worked closely with George Dollond, having moved into the Fleet Street premises. Most likely, generally Tulley served as an apprentice to Peter Dollond before starting out on his own. While Dollond is best known for the achromatic lenses under 4 in. in aperture, Tulley is credited with producing a lens as large as 6.8 in. in diameter – an enormous challenge for the day and age.

Tulley died in 1830, but the firm continued under the aegis of his two sons, William and Thomas. William Tulley was an innovator in his own right, credited with making the first achromatic microscope objective lens. In addition, Admiral W.H. Smyth used a 5.9-in. refractor made by the younger Tulley in 1828 to carry out extensive astronomical observations. These were eventually published as *A Cycle of Celestial Objects* in 1844, which served as the first-ever popular guidebook to the charms of the night sky and tailored for the adoring eyes of amateur astronomers.

Fig. 5.2 A Broadhurst Clarkson spotter c. 1900 (Image credit: Science Collectables)

Robert Mills acquired the business from the Tulley brothers in 1844 but eventually sold it to an enterprising telescope maker named Alexander Clarkson in 1873. Finally, in 1892, Clarkson was joined by Broadhurst as a partner. By this time the firm mainly concentrated on the production of telescopes for both nautical and astronomical use, though they also churned out microscopes to order (Fig. 5.2).

Unfortunately, the two partners never saw eye to eye and went their separate ways in 1908. Broadhurst then moved from Fleet Street to 63 Farringdon Road, naming the building 'Telescope House' and, some say facetiously, decided to trade under the name Broadhurst Clarkson & Co – both to trade on Clarkson's reputation and also to irritate him. Indeed, the original sign is still on the building today and is a listed London landmark.

Fortunately, the company continued to grow from strength to strength, not only retailing their own instruments but supplying other telescope manufacturers with drawn tube and castings. The firm also secured lucrative contracts with Her Majesty's War Office. Indeed, throughout the duration of World War I, the company had to hastily set up makeshift telescope factories in Watford, as well as further premises at 69 Fenchurch Street EC3 and 5 London Street EC3 (Figs. 5.3, 5.4, and 5.5).

This author spoke with London-based Gerald Morris, a former employee of Broadhurst Clarkson. Gerald joined the firm in 1966 and served as invaluable source of information concerning the day to day running of the business:

I was employed principally to make the brass tubing for the telescopes, turning the metal rolls using a mandrel and hand-polishing each tube before fitting the baffles and the object glasses. It was a thriving business back then, the company having received commissions from the army, navy. The lenses were ground, figured and polished in house. They were left uncoated, though, to preserve the best polish possible. Each lens was star tested using an artificial point source set up in the workshops. If it didn't give the right intra- and extra-focal pattern it was sent back for refiguring and/or polishing. While there was never much difference between an ordinary object glass and our best lenses, the latter were outfitted in

BROADHURST, CLARKSON & Co.'s
Famous Standard
Astronomical Telescopes

For Amateur and Professional.

Most people have no idea of the extraordinary number of stars visible through a small two or three-inch glass : not only stars, but *craters of the moon, spots on the sun, phases of Venus, Belts of Jupiter, rings of Saturn,* and a host of *double stars* and *nebulæ* are within easy range of one of these splendid instruments offered below. The lenses used in all models are warranted perfect in workmanship.

Every model is carefully tested and offered as the finest value possible.   Here is a small selection from the wide choice available.

2¼-inch glass telescope.   Brass lacquered body with astronomical eyepiece, but without stand   - £5 5 0

3-inch ' Starboy ' as illustrated. Main body of brass, rack and pinion for focussing, standard astronomical eyepiece (80x) with suncap and terrestrial eyepiece (40x) table stand, complete in polished pine case   -   -   -   £12 10 0

3-inch Telescope with 2 astronomical eyepieces (80x 160x) on tall garden tripod stand - £15 0 0

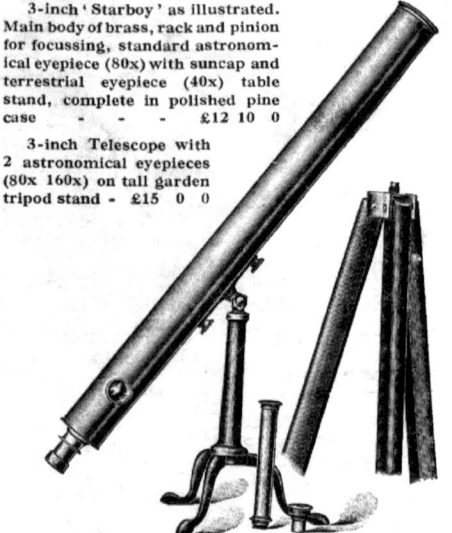

Many cheaper models also obtainable.

NOTE.—The makers can supply sets of lenses and parts to make your own telescope at home.   Prices from 10/-

Full descriptive catalogue free on application to Broadhurst, Clarkson & Co.

All applications should be sent to—

BROADHURST, CLARKSON & CO.,
c/o Messrs. Chas. Pool & Co., Ltd.,
180, Fleet Street, E.C.4.

Fig. 5.3 An advertisement for Broadhurst, Clarkson dating from the early twentieth century (Image credit: Phil Jaworek)

Fig. 5.4 Man about his work at Broadhurst Clarkson (Image credit: Steve Collingwood)

the company's deluxe telescope models, which usually ranged in aperture from 2.5 to 4 inches, with the occasional commission of a 6 inch instrument. The vast majority of our buyers hailed from the UK, but there were a few notable overseas customers, including a famous Sultan from the Middle East, who commissioned a small tabletop refractor of 2.5 inch focus the tube of which was made from rolled solid silver and after polishing encrusted with precious gemstones! Our smallest, portable telescopes came with a modest price tag of £40 and the modest popular model for amateurs was a 3-inch altazimuth mounted refractor that had a £150 price tag.

By the 1970s, business had begun to wane, due in large part to the rise of high-quality Japanese imports. It was simply more cost effective to order in the lenses than continue to make them at Telescope House, so the extra premises was closed and the entire operation returned to 63 Farringdon Road. Nonetheless production did not cease entirely, with Benjamin Martin's original draw bench still in use.

Around this time, Broadhurst had begun to buy mounts made by a telescope maker called 'Fullerscopes.' Rather amusingly, Broadhurst would have his engineers file off the name 'Fullerscopes' from the mounts he purchased, but Fuller countered by having the castings embossed so that his name could not be erased so easily (Fig. 5.6).

Fig. 5.5 Everyday work at Broadhurst, Clarkson & Co in the early 1970s (Image credit: Steve Collingwood)

Fig. 5.6 The newly produced telescope lenses laid out for inspection (Image credit: Steve Collingwood)

By 1973 the firm was on its knees under the weight of competition from much cheaper Japanese imports and the escalating cost of manufacturing telescopes by hand. It was at this stage that the company was purchased by Dudley Fuller, who brought with him his 'Fullerscopes' brand of telescopes and mountings and re-named the firm Broadhurst, Clarkson & Fuller Ltd. These products resonated nicely with the products manufactured at Telescope House. Dudley set about putting Telescope House back on the map – staffing the shop with experienced amateur astronomers and tailoring production to meet the varying budgets available to its customers.

Traditional telescope production still continued under the guidance of master craftsman such as the late Ernie Elliot, who had started with the firm during World War II and had since become the senior craftsman within the firm. Although demand for newly crafted traditional refractors diminished, Ernie was kept very busy renovating and restoring brass telescopes and microscopes for auction houses, antique dealers and collectors. Ernie continued to make instruments using the traditional techniques and tooling (some of it more than a 100 years old).

However, this was a market in rapid transition, and technology was winning the economic battle. Accordingly, in 1983 BC&F became the sole importer for Meade instruments in California as well as the UK importer for Celestron. The firm continued to grow and very soon passed over the Celestron dealership to concentrate on Meade with the advent of their GOTO telescopes. Traditional production continued, however, until approximately 1992, when Ernie retired and the old draw bench was donated to the Museum of Science and Industry in Birmingham.

By 2005 the company had finally outgrown 63 Farringdon Road, so it was time to expand. The company relocated to Tunbridge Wells in Kent, and the doors were finally closed to No. 63. The year 2009 saw the company relocate again, this time to Starborough Farm in Edenbridge. This was then followed by the opening on the new retail Telescope House in Lingfield in Surrey shortly after. Although the main business of BC&F remains the import, distribution and sale of modern telescopes and accessories, the company history still plays an important role in the firm's identity (Fig. 5.7).

Phil Jaworek, an amateur astronomer based in England and classic refractor fan is the lucky owner of a 1970s vintage 4-in. f/15 Fullerscopes refractor as well a fine 3-in. aperture Broadhurst-Clarkson Starboy refractor dating from the early twentieth century. Phil described his journey into the exciting new world of restoring and using these classic British scopes (Figs. 5.8, 5.9, 5.10, and 5.11).

For many years, I have taken a keen interest in antiques and collectables, regularly visiting fairs and auctions to see what bargains could be had. Unfortunately antique and vintage astro-gear is rarely seen, and when it is, the prices are usually, excuse the pun, astronomical. But in early 2010, I discovered an interesting 'scope and mount at a local auction house. I placed a tentative commission bid and then spent the next few days wondering if I should have bid higher. Amazingly, my bid won, and I became the proud owner of an early 3 inch Broadhurst Clarkson brass astronomical telescope together with a massive but unidentified equatorial mount. Later investigation revealed it to be a late nineteenth century Thomas Cooke & Sons of York Portable Equatorial Mount. The 3-inch brass telescope was a good example of a late nineteenth/early twentieth century 3-inch Broadhurst Clarkson Starboy refractor, but it looked lost on this massive mount.

Fig. 5.7 A 4-in. Broadhurst Clarkson (Image credit: Richard Day)

What that mount really needed was something bigger; preferably a similarly aged Cooke refractor. Unfortunately, these are rare and very expensive, thus an alternative was investigated. A few weeks later, after speaking with a local telescope manufacturer, I found he had a vintage 1979 4-inch F15 Fullerscopes Deluxe refractor for sale. These 'scopes, even though they were produced well into the 1980s and early 90s, were made in the style of the nineteenth century Cooke telescopes using traditional techniques with lacquered brass components, a perfect match for my mount.

The objective serial number on the 4-inch Fullerscopes puts it at 1979 when Henry Wildey was producing lenses for Dudley Fuller, I understand. Although I can't prove it, the quality of the objective suggests it may be one of his. The Deluxe scopes (which this is) had

Fig. 5.8 A handmade c. 1910 3-in. Broadhurst-Clarkson 'Starboy' refractor (Image credit: Phil Jaworek)

cherry--picked optics according to the BC&F catalogue of the day. That together with knowing Ernie Elliot put the telescope together makes this instrument special to me. I would like to see anyone with a Celestron or William Optics put names to the people who made their scopes

Due to the age of the mount and historical interest I did not wish to make any major modifications so I spent a few hours doing a gentle clean up and refurbishment. However, the 100+ year old leather straps were showing signs of age, and I could not trust them to support a heavy f/15 scope; they were duly replaced with a pair of Fullerscopes cast tube rings that came with the 'scope.

Fig. 5.9 A Fullerscopes 4-in. f/15 classic refractor (Image credit: Phil Jaworek)

Because the new 'scope was longer than the original 3-inch 'scope fitted to the mount I found operating the declination shaft lock difficult when viewing through the eyepiece. Therefore I manufactured an extended declination axis lock screw from brass using the illustrations in the Cooke catalogue as a guide. This modification provides convenient control of the declination axis close to the eyepiece end of the scope.

I have a few accessories, including a 1920s brass B&C Star Diagonal, which I modified to take 1.25 inch eyepieces and that looks a treat, but I've also converted a William Optics dielectric diagonal to brass, and this not only looks the part but works superbly with it. Optically, the 4″ f/15 it is superb and my instrument of choice for looking at the Moon, bright planets and double stars. It has fulfilled every expectation I had for it.

As we shall see in a later chapter, Telescope House still has the capacity to restore fine old instruments to good working health. Indeed, we shall explore how two classic refractors, owned by a very famous astronomical evangelist, were given a new lease of life in the capable hands of its technicians.

Fig. 5.10 Moonstruck! The 4-in. Fullerscopes captures some awesome shots of the barren lunar regolith (Image credit: Phil Jaworek)

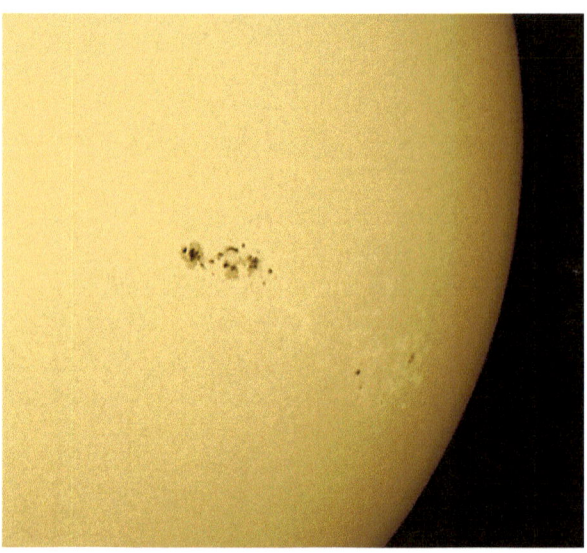

Fig. 5.11 Top performer – the 4-in. f/15 Fullerscopes images the solar disk (Image credit: Phil Jaworek)

Chapter 6

The 60 mm Brigade

It's true; the long focus 60 mm (2.4-in.) refractor has developed a bit of a Jekyll and Hyde personality among amateur astronomers over the years. On the one hand, for some it should be banished forever to the fires of Hades for the poor optics and mechanics it served up to an army of unsuspecting newcomers unfortunate enough to end up with one of those junk (read, discount department store) 'scopes that have flooded shopping malls and camera shops over the last 20 years. These 'things' probably did more to discourage curious beginners interested in the hobby than advance the cause.

However, for others lucky enough to have acquired earlier models, nothing could be further from the truth. Indeed, if you start talking to a broad cross-section of amateurs, you see a different picture of the long focus mini-scope emerge. Many Nordic and Eastern European amateurs, for example, not only received their first formal training in telescopic astronomy with refractors of this size (especially the superlative Zeiss Telementor), but for the majority, it was the only decent 'scope readily available to slake their thirst for heavenly light. Unfortunately, for the most part, their testimonies remain silent, their voices as yet unheard across the Internet.

Others blessed with the means to own larger 'scopes with good optics have dismissed even the best performers in this genre as rather aperture challenged and thus not likely to excite a child or an inexperienced adult. But having re-examined the question, the latter reaction seems to be more a logical construct than a reality.

For this study, the author has attempted to put the ugly attitude of 'aperture elitism' to one side and will compare the views through two little achromatic refractors from memory lane – a late 1950s Swift model #838 50 mm F/14, and a late 1970s Tasco model # 14TE 60 mm F/15. Finally, a few impressions of a Tasco model #58 60mmF/11.4, marketed in the mid 1980s.

N. English, *Classic Telescopes: A Guide to Collecting, Restoring, and Using Telescopes of Yesteryear*, Patrick Moore's Practical Astronomy Series, DOI 10.1007/978-1-4614-4424-4_6, © Springer Science+Business Media New York 2013

A few questions first came to mind:

1. Were there any big differences in optical quality in these 'scopes separated by the decades? Was a 'scope used in my father's time better than a scope I might have enjoyed as a teenager?
2. How did the mechanics of the package in both instruments differ from each other?
3. What do the answers to questions 1 and 2 mean for our hobby in general?

First Impressions

The Tasco 'scope arrived in its original box. When I unwrapped the packaging I was amused to see that though the paintwork had faded on the cardboard outer casing, my eyes met with incredible images of nebulosities – the celestial Crab (M1) and the Orion Nebula (M42), as would be imaged by large 'scopes. Then I noticed a caption on the side of the box boldly proclaiming, *"Today's Hobby, Tomorrow's Profession."* OK. But all jokes aside, what did I get for my $100? (Fig. 6.1).

The contents included the OTA, equatorial head and small electric RA motor, tripod legs, two Huygenian eyepieces (26 and 6 mm), a shade for solar projection (Solarama), a conventional 2× Barlow lens and a curious "zoom eyepiece" that enabled me to use the 26 mm ocular at a variety of focal lengths down to 7.5 mm. Zoom eyepiece? Eh? Zoom Barlow more like!

Fig. 6.1 The c. 1978 Tasco 14TE arrives in a polystyrene-lined cardboard box (Image by the author)

Fig. 6.2 The late 1950s 50 mm Swift arrives in a custom-made wooden box (Image by the author)

Contrast that to the presentation of the smaller Swift refractor. Unlike the 14TE, this baby arrived in a custom wooden case, neatly compartmentalizing all the goodies, which included the equatorial mount and wooden tripod, a terrestrial viewer and a Barlow lens. The owner also kindly threw in a couple of high quality 0.965-in. Carton Plossls. It was a most satisfying presentation. Everything had its place. Neat! (Fig. 6.2).

Assembly

The 14TE, which appeared very lightly used for its age, was easily assembled from scratch in less than ten minutes, with no tools. Its all metal, die cast head has a strong, workmanlike appearance with good mechanics. Displaying the Circle T logo, the mount is mainly of die cast metal construction, with slow motion controls in both axes that allows the 'scope to be moved smoothly in both Right Ascension (RA) and Declination (Dec). It was heavily greased, however, which sadly reminded me of the current crop of entry-level mounts (Fig. 6.3).

In contrast, the Swift was set up in less than half that time, a clear reflection of the superior engineering that went into it. For example, with the Swift 'scope, provision is made for removing the equatorial head from the base of the mount with a single, well-machined, bolt. Unlike the 14TE, there is no hint of the mechanics that enables the RA and Dec axes to move, yet move they do – without loosening or turning a single knob, and all with the gracefulness of a Swiss watch (Fig. 6.4).

Fig. 6.3 The Tasco 14TE mount is light but sturdy. It is heavily greased, and you can see it (Image by the author)

Fig. 6.4 The over-engineered Swift mount moves with the precision of a Swiss watch. Move on folks! No grease to be seen here (Image by the author)

Fig. 6.5 Both 'scopes are a breeze to set up and transport into the field. Ultraportable these are! (Image by the author)

Once assembled, there was no need to take either 'scope apart. I just removed the optical tubes from their cradles and parked them neatly away in the corner of my office. Most anyone could probably carry these mounts easily with one hand, so they're ready to go in a jiffy (Fig. 6.5).

The wooden tripod accompanying the Swift was strong yet lightweight. Apart from natural fading and chipping away of the varnish, it still presented in fine working condition. Unfortunately, I can't say the same about the Tasco tripod. When I fully extended the legs, I discovered, to my dismay, that they were not rigidly affixed and actually had a dangerous amount of wobble to them. The problem was remedied by pushing the legs back in a few inches (Fig. 6.6).

A Closer Look

The Tasco 14TE arrived in remarkably good condition for a 'scope that saw first light over 30 years ago. The optics looked pristine. Ditto the Swift, though its objective could do with a clean up some time soon. Alas, the focuser knobs on this telescope are plastic, a subtle departure from the all metal focusers of earlier 60 mm Tasco models. As expected, there wasn't any plastic on the little Swift (Figs. 6.7 and 6.8).

Fig. 6.6 Check out the bandy legs on the 14TE. Not nice! (Image by the author)

Fig. 6.7 The simple rack and pinion focuser on the 14TE. Check out the plastic focus knobs (Image by the author)

Fig. 6.8 No plastic to be seen on the Swift focuser, or anywhere else, for that matter (Image by the author)

Interestingly, despite being longer and wider than the Swift, the 14TE Optical Tube Assembly (OTA) weighed slightly less than the former (about 2.5 lb). The Swift is one chunky 50 mm 'scope! Both single speed, rack and pinion focusers on these telescopes work extremely well, though the Swift had a fit and feel that was, irrationally or otherwise, more pleasurable to use. I'd rate it up there with the fine focuser on TeleVue doublets.

I also had a look inside the tubes to see how they dealt with suppressing stray light. The metallic dew shields on both instruments unscrewed to reveal two nicely machined lens cells. The inside of the 14TE tube assembly showed a matt blackened interior with little sign that it had deteriorated over the decades. It had what appears to be a single knife-edge baffle placed well down the tube and only a little bit ahead of the focus draw tube. The Swift had a more 'conventional' design, with a nicely blackened tube wall, together with a pair of light baffles, one placed about 6 in. from the objective and another located a few inches further back. The front baffle looked like it could do with a new coat of paint, though.

Examining the finders, I was in for a shock! It was clear that, with the Tasco at least, the rot had set in. It was poorly made, with a central obstruction so absurdly large that it rendered it next to useless. Even at its sharpest focus (it took an eternity to get there, as it was so stiff), the views were dim and fuzzy. You can understand why I took a scunner to it! (Fig. 6.9).

Contrast that to the fantastic counterpart supplied with the 1950s Swift. The tiny, coated objective was unobstructed and housed inside a nicely machined metal OTA (Fig. 6.10).

Fig. 6.9 Call that a finder on the 14TE? Scunnered! (Image by the author)

The Swift had other goodies that I found neat to use, most noticeable of which was the terrestrial viewer that folds the light path in such a way as to provide a properly orientated, upright view (Fig. 6.11).

The objectives in both these small telescopes are housed in very well machined cells. Neither has provision to easily collimate the lens, though. The sexagenarian Swift is outfitted with a cute little air-spaced objective. Interestingly, I learned that the objectives for these early Swift 'scopes were made either by Takahashi or H. O. C. The anti-reflection coatings looked as though they were applied yesterday.

The 14TE's objective was more elusive, with more subdued coatings and no sign of an air space between the lenses. Was this a contact/cemented doublet? Intrigued, I unscrewed the retainer to find that the crown and flint elements were actually separated by a thin ring spanning the circumference of the objective, so air spaced it is, too (Figs. 6.12 and 6.13).

Optics and All That

As with all telescopes that pass my way, I put these puppies through their paces by day, as well as under the stars. Both 'scopes came with prism diagonals that accept 0.965-in. oculars. Attempting to level the playing field a little, I elected instead to use a high quality Takahashi prism diagonal in both 'scopes and ordered up a 0.965–1.25-in. adaptor £15 expenditure) to enable me to use my orthoscopic eye-pieces. That's when I hit my first hurdle (Fig. 6.14).

Fig. 6.10 No such thing with the Swift finder. High quality, unobstructed optics (Image by the author)

Fig. 6.11 Another nice Swift touch – a high quality terrestrial viewer can be added for day-time observing (Image by the author)

Fig. 6.12 The 14TE has an air-spaced doublet objective, though it's hard to tell until you take a closer look inside. MgF$_2$ coatings were applied to inner and outer surfaces (Image by the author)

Fig. 6.13 The dusty Swift had an air-spaced doublet with all four surfaces coated with MgF$_2$ (Image by the author)

Fig. 6.14 An inexpensive adaptor enables me to use a high quality prism diagonal and my favorite 1.25-in. eyepieces (Image by the author)

The Takahashi prism diagonal allowed me to achieve infinity focus in the 14TE without much difficulty. However, it was a no-go with the Swift. Anything beyond about 30 m was out of bounds with that set up. No problem. I settled on testing this 'scope using its dedicated 0.965-in. prism diagonal, together with a few nice Carton Plossls thrown in by the 'scope's owner. Alternatively, I could also make use of the adapter to allow my 1.25-in. orthos to be used if and when required (Figs. 6.15 and 6.16).

A quick test with a Cheshire eyepiece showed the Swift to be in perfect collimation. The Tasco 14TE was a bit off, but nothing to worry about. After all, I had learned from experience that such long focus 'scopes are very forgiving to misalignments. Nonetheless, I had a go at tweaking collimation by loosening the retaining ring and gently strumming the outside of the cell to see if the lenses would 'settle' into a slightly better position. Attempting this a few times finally gave results that were a significant improvement over the original settings (Fig. 6.17).

Inserting my 18 mm ortho, I was greeted with delightful high contrast images of my rural hinterland. Observing the variegated hues of autumnal leaves against a bright sky background is one of the harshest tests you can subject a refracting telescope to. Short and medium focal length achromats reveal the color purple all too easily when pushed to moderate and high powers at the eyepiece. Short focus ED doublets are more subtle, but still follow suit at higher powers. I also discovered a glitch with the 20 mm Carton Plossl. It displayed a noticeable amount of lateral color in comparison to the 1.25-in. orthos I had been using with the 14TE.

Fig. 6.15 Check out this little piece of engineering magic. The Swift comes with a precision made prism diagonal and a lovely helical eyepiece grip (Image by the author)

Fig. 6.16 I used some nice 0.965-in. Carton eyepieces with the Swift (Image by the author)

Because these little refractors run out of light very quickly, I had to wait for a cold, bright, sunny day to push the optics hard. I inserted a 6 mm ortho into the Tasco and was greeted with images that were comely and color free at 150×. Ditto

Fig. 6.17 A good way to check initial collimation – just add an inexpensive Cheshire eyepiece (Image by the author)

for the Swift at 117×. Contrast and definition were excellent in both. Tiny droplets of water were resolved into perfect spheres, reflecting white light. Nor did I see any color fringing in the feathers of a curious magpie that frequents the conifer trees near my home.

Readers shouldn't be surprised by this. After all, these 'scopes have chromatic aberration (CA) indices (found by dividing the focal ratio of your 'scope by its aperture in inches) of 6.25 and 7.0, respectively. As an additional test, I set up another 'scope, a red model # 58 T (60 mm F/11.4) Tasco alongside the others. Though the images were essentially color free at low and moderate powers, I could indeed detect some false color on the same targets when I pushed the magnification to 117×. Interestingly, according to Conrady, a CA index of 5 is considered the minimum for semi-apochromatic performance, and the 58 T just falls short of that minimum. Furthermore, the images in the 58 T were definitely a shade less 'punchy' than the older 'scopes when viewing highly 'texturized' objects, such as a tree trunk examined in the cold light of day (Fig. 6.18).

Another unsung virtue of these little refractors is their enormous back focus. Racking the focuser tube outwards, I was able to use the 14TE as a long range

Fig. 6.18 Where's that nosy magpie? (Image by the author)

microscope. I managed to get the 'scope to focus (with a small extension tube) on one of my garden hedges just 7 yards away. The activity is highly addictive. I struck Safari gold when I chanced upon on a female Orb spider at 150× spinning a deadly web, her elegant tan colors contrasting beautifully with the white spots that speckled her back (Fig. 6.19).

Star Testing

After receiving both 'scopes, the predictable happened; I had to endure a week of howling winds and unrelenting rain before finally getting a chance to star test both the Swift and 14TE. With these tiny apertures, you need as much light as possible to get good information about your optics. For that, I turned to the brightest star best positioned at the time. With Vega high in the northwestern sky, I first looked at a high power, in-focus image (117× and 150×, respectively).

Fig. 6.19 The generous back focus on these small refractors makes them ideal as long distance microscopes (Image by the author)

The Airy disks were clean and crisp in both instruments, surrounded by a perfectly circular first diffraction ring. There was no sign of the 'fish gills' suggestive of coma. Furthermore, they were presented in sharp, high contrast, with no detectable violet halo. The first diffraction ring was more prominent in the Tasco than in the Swift, a portent of minor spherical aberration. Racking the 'scope inside and outside focus in the Swift yielded well nigh textbook symmetry in the pattern of the diffraction rings. Indeed, the only way I could tell one side from the other was by the color of the rim – green outside and faint aniline inside.

Repeating the same tests with the 14TE, I detected a very slight asymmetry in the intra- and extra-focal patterns. They were beautifully rendered inside focus but a tad less defined outside. Repeating the same tests over a number of nights showed much the same result. Consulting Suiter's book on star testing, I judged the 14TE lens to be slightly under-corrected. For the record, I also detected the merest trace of astigmatism when the 14TE was pushed to powers beyond 150× – a consequence perhaps of the objective's slight mis-alignment with the eyepiece. That said, I wasn't at all bothered. It made no material difference to the in-focus images. Stars remained sharp and pinpoint all the way across the field of view in both 'scopes, suggesting little in the way of field curvature.

Bright stars are also great for assessing unwanted glare. Pointing both 'scopes at Sirius in the small hours of the morning, I compared the low power in both the Swift and 14TE (35× and 38×, respectively). Running back and forth between them, I formed the definite impression that the Swift had slightly more contrast than the Tasco. This could be attributed to the better coatings applied to the Swift in comparison to the 14TE. All in all, I was thrilled with both, but if push came to shove I'd have to award the Swift an A+, and the 14TE an A-.

Ad Astra

With the testing over, I was free to go hunt celestial real estate. My first target was post-opposition Jupiter. For this test, I charged the Swift and 14TE with powers of 80× and 100×, respectively. Despite the planet's low altitude in the sky, both 'scopes rendered sharply focused images of the giant planet with no detectable violet halo. I checked the color correction by setting up my 3-in. F/6.3 doublet apochromat along side. Though understandably putting up a brighter image and resolving finer details, I judged their color correction to be quite comparable.

Like a spacecraft test image from a distant vantage, the Jovian globe was seen to have at least two fawn-colored bands, and on one occasion, I was able to just make out the Great Red Spot. Seeing this structure in such a small instrument is a challenge at the best of times. I wonder if the temporary disappearance of the planet's South Equatorial Belt made the difference here. Whatever the reason, the whole experience left me with the unmistakable impression that if I trained my eyes real well, I might have divined more from these images. Certainly, keener-eyed folk would surely be able to prize even more detail from either instrument.

The first-quarter Moon was an absolute joy to behold in these little telescopes. Both presented the lunar regolith in its true pastels. One look through the Swift at 80× reminded me why I've never quite had the same fondness for lunar observing that I have for the idle simplicity of looking at double stars. Luna, even in these tiny refractors, presents me with information overload! After all, it pays to remember that even a 2-in. glass is capable of revealing craters of the order of 4 miles across and can just reveal lunar clefts as narrow as 400 or 500 yards in width.

At a glance, I could make out a suite of different selonographic terrain – vast swathes of smooth Maria, fields of craters and long and winding valleys. But for me, it is the scrutiny of the wondrous Apennine Mountain range that tops the whole lunar experience. The changing aspects of the shadows cast by their tallest peaks – some of which soar 15,000 ft above the surrounding plains – was an awe-inspiring sight with these little spyglasses. The Southern Highlands, battered and bleak, presented endless opportunity for studying crater morphology.

When pressed to the highest powers I could squeeze out of these small refractors, there was, so far as I could make out, no false color along crater rims. Here were two little achromats, I thought, performing just like their younger siblings, the apochromats.

Truth be told, having become accustomed to the views served up by larger instruments, I had almost forgotten just how good these little telescopes were, particularly when employed to the task of resolving selected double stars. As I discovered, these telescopes present some of the most enchanting views of distant suns you can get.

Bright stars simply don't present as pinpoints of light in these small refractors so much as they show Airy 'globes' (a consequence of the telescopes' small aperture) surrounded on the good nights by a single, perfectly formed diffraction ring. Indeed, the stellar images in these instruments may be as close to theoretical perfection as ever you're going to get. Showpiece systems, such as snow white Mizar and Alcor in the handle of the Big Dipper asterism, emerald and gold Albireo in Cygnus and the lovely orange pair known as 61 Cygni (Piazzi's flying star) were gorgeous and easy at low power in both the Swift and the Tasco.

Swinging the 'scopes over to the eastern sky, I was greeted by the two brightest of Gemini's suns, Castor and Pollux, loitering with intent above the tree lines. The mortal luminary, Pollux, has a clandestine prize, only revealed by charging the telescopes with fairly high powers. Castor, both telescopes show, is not singular. The two brightest members of this multiple-star system – Yin and Yang to the ancient Chinese – are shamelessly unveiled in their virginal white and citrus lemon. From our cozy vantage, 52 light years distant, the pair looks serene, as if frozen in space and time, yet careful observations over many decades with instruments as diminutive as these can show they are compelled to move, each member orbiting the system's common center of gravity and completing one lap in 450 years or so.

Theoretically, the 50 mm Swift should resolve equally bright sixth magnitude pairs with a separation of 2.3 s of arc. The 60 mm Tasco ought to do considerably better, having a Dawes limit of 1.9 arc seconds. The famous double double in Lyra (Epsilon 1 and 2 Lyrae) provides an excellent test for these 'scopes (separations of 2.1 and 2.4 arc seconds, respectively). Though I tried on a few occasions, the little Swift really struggled with this system. The best I could get on my nights out with it was two elongated stellar pairs.

The 14TE, with its extra 10 mm of aperture, fared much better. On calm nights I was able to coax a nice separation from both pairs at 150×. On a night of excellent seeing, the charming couples, lying nearly at right angles to each other, were framed as four perfectly formed disks of luminous energy, surrounded by a tranquil diffraction ring – an enchanting moment indeed!

Next, I turned both 'scopes on Izar (Epsilon Bootis), now, sadly, sinking fast into the northwestern sky. To get the best views, I observed immediately after sunset, when a brief window of opportunity was granted to me, during which time the air can be calm and stable for a spell. Though I had to try a few times, my patience paid off by glimpsing the Neptune blue orb nestled right up against the orange primary. Needless to say, it was a sight for sore eyes! Once again, aperture won the day. I have yet to resolve *Pulcherimma* in the 50 mm Swift, though I'm holding out for a magical night next spring, when the system is better positioned for observation.

Out for more challenges, I tried splitting the faint companion to Polaris with both 'scopes. An instrument this size requires good optics and good seeing in equal measure to pull off such a feat. That's not because Polaris B is especially close to its primary (at 18″ arc seconds, it isn't) but rather because there is such an enormous chasm between their apparent magnitudes. Polaris is a second magnitude star, its companion a miserable ninth Charging the 60 mm F/15 Tasco with a low power of 38×, I could detect its ghostly glow intermittently on average nights of seeing

Fig. 6.20 Here's a book that will help you find your way around the sky with these small 'scopes (Image by the author)

and transparency. Throwing more magnification at it seemed to work, up to a point. At 72×, for example, the tiny speck of light accompanying the North Star was more comfortably held. In contrast, there was no sign of the companion in the smaller Swift at magnifications ranging from 35× through 117×, at least during the time the telescope was entrusted to me.

Empowered with a good star atlas, there are literally hundreds of double stars that will keep you on your toes with these small aperture 'scopes. Sometime soon, I plan to look for Rigel's elusive companion and resolve all three components of Iota Cassiopeiae simultaneously. So many targets, so little time! (Fig. 6.20).

When equipped with modern wide-angle eyepieces, these long focus mini-refractors can actually deliver a respectably wide field of view. To that end, I inserted my 24 mm super-wide angle eyepiece into the 14TE. Such a combination yielded 38× and a 1.8° field of view. OK, it wasn't as wide as could be achieved using the currently available Takahashi FS60C and/or TeleVue 60, but it is wide enough to see the majority of deep sky objects in their entirety. Accordingly, I chased down the double cluster in Perseus, the glories of the Pleiades and Hyades in Taurus and the Great Nebula in Orion (M42). One of my favorites, the open cluster M35, anchored to the northern foot of Gemini, was well framed, though admittedly, owing to the open cluster's abundance of fairly dim stars, I was quickly pining for more light-gathering power (Fig. 6.21).

Clearly, these were not the bogus 'scopes so derided by the astronomical cogno-scenti! They performed well beyond my expectations! So why aren't more of us wooed by the charms of the humble 2-in. glass? For many, that enchanted state was

Fig. 6.21 The 14TE kitted out with a modern, wide-angle eyepiece works surprisingly well as a rich field 'scope (Image by the author)

shattered utterly and forever sometime beginning in the mid-1980s, when the shifting sands of corporate beliefs led to the slow relegation of the 60 mm refractor to the trash heap of amateur astronomy.

Formed in 1954, Tasco (or more accurately, Tanross Supply Co.) commissioned its telescopes from various suppliers in Japan. The best glass, so this author has been reliably informed, came from Astro Optical Industries Co., Ltd. (Royal brand), but later Tasco contracted the work to other Japanese companies (Towa, especially) just like a proliferation of other instruments marketed by Swift, Unitron, Jaegers, Sears, Jason, Pentax and Mayflower, to name but a few. In Britain, many fine Japanese objective lenses found their way into telescopes marketed by Hilkin and GreenKat (Fig. 6.22).

Some companies, like Unitron, quit while they were ahead, so to speak. They wrapped up the business of sculpting long focus refractors in 1992. Curiously, research shows that the optical quality of this highly regarded brand actually varied quite a bit. Some are just OK, while others are exemplary.

Tasco didn't give up, though. They developed a different – and some would say aggressive – business model, one that greatly increased productivity but only by compromising on optical and mechanical quality. New optical firms, with their extremely low labor costs from Korea, China, Taiwan, Hong Kong and the Philippines, competed for the contractual rights to outfit Tasco with their optics. Tasco, it seems, sold out to the lowest bidder. Fueled mostly, it seems, by the prospect of big profit margins, the company peddled more 'scopes than all other brands put together!

Fig. 6.22 The fine 60 mm f/13.5 Hilkin achromat looks very similar to those sold under the name of Swift (Image by the author)

For me, the 14TE represents a very special Tasco. It's quite clearly a hybrid 'scope. On the one hand, the Towa objective is very good indeed. That said, it is abundantly clear that the company was beginning to cut corners in outfitting the 'scope with its accessories. And although the mount is nicely machined metal, it is quite simply in a different (read lower) league to the craftsmanship evident in the earlier Swift model. If you take a look at the 7TE, the antecedent of the 14TE, you'll get optics delivered in a collimatable cell, an even better mount, a high quality finder and no plastic to be found anywhere (Fig. 6.23).

These sentiments were consolidated after putting a mid-1980s 60 mm Tasco through its paces. Examining a red-tube model 58TE (60 mm F=700 mm). Right off the bat, I noted the objective is mounted in a cheap plastic lens cell with screws that protrude into the optical train (Figs. 6.24 and 6.25).

What is more, instead of the finely machined prism diagonal in a nice metal housing that accompanied the 1950s Swift and the 1970s Tasco 14TE, the 58 T's counterpart is a cheap plastic model. Even after upgrading it and inserting some high quality orthos, this 'scope star tested badly, showing significantly more astigmatism and spherical aberration. And while the images were acceptable at low and medium powers, high magnification views of the Moon and Jupiter were noticeably less sharp in this 'scope than either of the older 'scopes (Fig. 6.26).

Fig. 6.23 Check out the attention to detail in this Tasco model 7TE, complete with fully collimatable objective (Image credit: Richard Day)

Fig. 6.24 Check out this botched job on the Tasco 58T! Note the plastic lens cell and protruding screw (Image by the author)

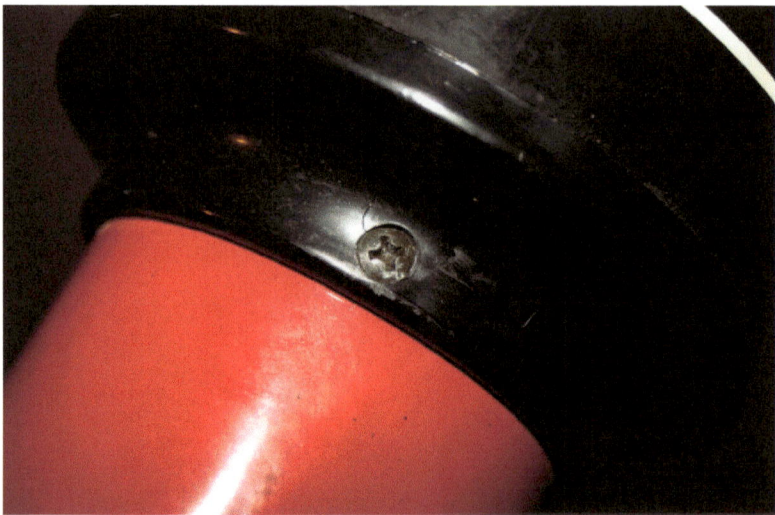

Fig. 6.25 Instead of a threaded lens cell, the optics of the 58T is strictly out of bounds with these screws (Image by the author)

Fig. 6.26 Prismatic junk on the Tasco 58T (Image by the author)

Looking into Space

So, which of the three 'scopes did I prefer? From a completely utilitarian perspective, the 14TE would have to get my vote. Though not quite as good optically as the Swift, it does enjoy 20% more resolution and gathers up 44% more light than the Swift and so easily wins under the stars. The Swift is an heirloom quality instrument designed for use during the day as well as by night. It is, in effect, a complete, mobile observatory/spotting station. The 14TE is more Spartan, intended to be used for astronomical applications primarily, but as I've shown, is equally adept at viewing objects in the daytime.

Would this author recommend these 'scopes to a newly minted amateur astronomer? In a word, no. Their needs would be better sated with a good 6 or 8-in. Dobsonian or a 4-in. refractor. Do I think they have any utility for the battle-hardened astronomer who has "seen it all"? Most definitely, yes! Here's why. Galileo saw and correctly interpreted the poor images served up by his tiny refractors. How much more would he have elucidated concerning the starry archipelago, were he to look through one of these instruments? The beauty of it all is that we'll never know for sure. But it's clear he wouldn't be drawing Saturn with plug ears!

If you have a spiritual side, then feast on these little telescopes for a while. Observing through a high quality 60 mm refractor (modern or classic) is great nourishment for the uninspired, especially if you happen to own a larger instrument! They don't give the kind of instant gratification that you'd get from a larger instrument, but the high power views they deliver on selected objects are vastly superior to many starter 'scopes currently on the market (especially of those ubiquitous 3–4-in. tabletop reflectors).

Three words summarize these 'scopes: faithful, pure and tranquil. Because these little telescopes are so immune to the distorting influence of the atmosphere, they almost always behave as if it were not there at all. Even modest increments in aperture create strikingly different impressions at the eyepiece. Comparing the views in a 3 or 4-in. glass to these small telescopes will easily convince you of that. When you use these 'scopes, it's like being in outer space!

In fact, these comparatively tiny 'scopes from the days of yore both represent, in miniature, the quintessence of what a refractor ought to be; it's that killer combination of uncontrived simplicity of design, high quality, long focal length optics, sound mechanics, easy portability and reassuring ruggedness that will forever make them endearing. They are two little belters from a unique period in history, when pride was taken in the construction of small telescopes. Then something bad happened. And the rest, as they say, is history – a beginning and an end.

Battle of 60 mm Classics

John Nanson, a gifted and evangelical double star observer from Oregon, kindly took the time to describe his adventures with several high quality 60 mm refractors that came his way in recent years and an assessment of their performance under dark, coastal skies. John has also made his own 60 mm refractors from high quality Japanese Carton lenses. As well as enjoying the views through his larger refractors, John took the time to describe his experiences of a few high quality classic 60s, and in particular an old Tasco 7TE and Carton 60 mm with a focal length of one meter.

How does the Carton 60/1,000 lens compare to the 60/1,000 lens in the Tasco 7 T-E? Now I'm really not sure what the age of my Tasco 'scope is, but I know the lens in it is sharp, crisp, and capable of resolving difficult doubles, so it's definitely not one of the later poor-quality versions. And the Carton 60/1,000 lens performs just as well.

Could one of the two lenses have an edge of some kind over the other one? The question deserves a good answer – so I set out to provide one. My approach was to put the two 'scopes on a Universal Astronomics DoubleStar mount, which is an alt-az mount built to handle two 'scopes side by side. I used the same kind of diagonal in each 'scope, as well as the same eyepiece. The diagonals are two identical older 1 ¼ inch tube types that I believe were made by Vixen. The eyepieces I used were two 17 mm Celestron Plössls, which provide very sharp views of doubles and, for these two 'scopes, a magnification of 59×. Because there might be small differences in performance between the two diagonals and two eyepieces, I swapped them back and forth between the two 'scopes during the test.

The first stop is Polaris, a mere 465 light years away. First, the data: magnitudes of 2.0 and 8.2, and a separation of 18.4". Now the Polaris primary seemed to be just a slight bit smaller in the Tasco. I also noticed a color difference. In the Carton, the primary appeared a bit less yellow and whiter, which seemed to have the effect of making it appear brighter than the image in the Tasco. I could just pick the secondary out with each 'scope, but there was some dampness in the air that made it very difficult to see it.

The next move was over to Eta Cassiopiae, a bit closer to home at a mere 19.4 light years. This one has magnitudes of 3.5 and 7.4, which are 13.0" apart. Even though the two stars are closer, the smaller difference in magnitudes between the two stars make it appear similar to Polaris – and that makes it a good one for confirming the differences I had just observed between the two 'scopes.

Once again, I noticed the primary was brighter in the Carton, and again there was a slight color difference – this time I could see a bit more yellow in the Carton. But what I also noticed was the secondary was slightly farther away from the primary in the Carton lens than in the Tasco.

Now this really puzzled me. I thought I had detected that same thing with the Polaris secondary, but the moisture made it so hard to see that faint point of light. This time, there was no question – there was a definite difference. In fact, as I looked closer, I realized there was a difference in image scale. The images of the primary and secondary were just a slight bit larger in the Carton, which was why the distance between the two stars seemed just a slight bit greater. I kept going back and forth between the two views. Finally I swapped the diagonal/eyepiece combinations between the two 'scopes, compared again, and found the views were identical.

I moved on to Iota Cassiopiae, a beautiful triple, with stars of 4.6, 6.9, and 9.0 magnitudes at distances of 2.9" and 7.3". The closer companion is the tough one here, and I was able to pick it out with both 'scopes. The distances were too close to reach any conclusions about image scale, though. Next stop was Rigel – magnitudes of 0.1 and 6.8, separated by 9.5" – which is always a good test of 60 mm optics. Again, both 'scopes split it with poise. But it was obvious that the primary was a bit brighter in the Carton.

Since I was near it, I swung the mount over to M42 – the Orion Nebula – and just stared at it for several minutes. When I started comparing, I could see a definite difference in contrast -- it was much better in the Carton. More of the nebula was visible, and the four stars in the Trapezium were just a bit brighter. North of Orion is NGC 1977, also known as the Running Man Nebula, and in the Carton the nebula was easily seen. In the Tasco, the nebula could be seen, but it wasn't quite as apparent.

So now I moved over to Beta Monocerotis, which is east of Orion. This one is an absolutely fantastic triple star. The primary, A, is a 4.6 magnitude star at a distance of 7.1" from B to C, which is a very close pair separated by 2.9" with magnitudes of 5.0 and 5.3. All three of these stars appear white to me, and when you see these three bright white stars so close together, it is stunning.

But I didn't really expect to split the B-C pair with either of these two 'scopes with the 59× contributed by the 17 mm eyepieces. And boy, was I surprised.

Now the split of B and C was amazing … and mesmerizing. I kept going back and forth between the two 'scopes, but the view in the Tasco kept pulling me back. And suddenly it struck me that these three stars were really *smaller and closer together in it. In fact, the 2.9" separation of B and C was clearly a bit larger in the Carton and a bit tighter in the Tasco. To be more precise, those two stars were so small and so close together in the Tasco that it was a miracle they could be seen as separate at all – which was certainly a testament to the quality of the Tasco lens.*

Finally, it was all starting to make some sense. The three stars of the ravishing Beta Monocerotis were so close together and so similar in brightness that it was much easier to see the differences between the two lenses. Polaris was the first clue, and Eta Cass left me with the realization that something really was different between the two lenses. So I went to bed and thought about it. In the morning I pulled out the tape measure.

When I brought the two 'scopes in from the observing session, I had left the focal positions exactly where they were when I was looking at Beta Monocerotis. So I measured from the center of the lens of each 'scope to the center of the diagonal, wrote down the numbers, and then measured the short distance from the center of the diagonal to the top of it. And I got a focal length for the Carton of 1,000 mm. The Tasco came in at 970 mm.

What I had seen was a difference in image scale caused by two different focal lengths. And slightly differing magnifications, but it was mainly the focal lengths that were the culprit. The true focal length of the Tasco comes in at f16.17, not f16.7, and the 17 mm Celestron Plössl was giving me 57×, not 59× in that 'scope. As the focal length of a given aperture increases, the image size increases with it, and that was what I kept bumping into all night.

And that brings us to the differences in contrast between the two 'scopes that I first noticed on M42. When I was looking at Beta Monocerotis, I noticed the sky background was a bit lighter in the Tasco. Both tubes are baffled similarly and are flat black inside, but the Carton 'scope I put together has a lens hood that I covered with jet black flocking paper on the inside. The inside of the lens hood on the Tasco is almost gray, not black, so my plans are to put some flocking paper on it and see if that corrects the problem. I don't think the contrast difference is a lens issue at all, especially after seeing that very delicate split of the Beta-BC pair.

And so, there you have it. Two 60 mm 'scopes, two lens of slightly different focal lengths, and a short tour of what I see every night I use these long white metal tubes of joy.

Can't see anything in a 60 mm 'scope, you say? Not a chance!

John Nanson's enthusiasm about these small 'scopes is positively infectious, don't you think? Indeed, he's a first rate double star observer, writing prolifically on their many virtues. Best of all, he continues to demonstrate to the elitists among us that one can have great fun under the stars with modest equipment.

A Twenty-First Century Revival?

In recent times, there have been signs that these small classic 'scopes are making a real comeback among a growing number of amateur astronomers – and not just classicists. Sheldon Faworski, formerly the owner of Apogee Inc, still sells high quality Carton objectives in 60 mm and 80 mm apertures in a variety of focal lengths. In fact, he also sells tubes and focusers to mount the optics to boot. Many ATMers have already built their own fine examples.

One British professional telescope maker, Richard Day, proprietor of Skylight Telescopes, London, has designed what is arguably one of the most beautiful of all classic 60mms, embodied in the Skylight f/15 m, the prototype of which this author had a chance to evaluate.

The Skylight f/15 m sports a top quality 60 mm (2.4-in.) Japanese objective with a focal length of 1,000 mm. The lens has a single layer magnesium fluoride coating and is housed in a carefully designed metal cell that can be collimated by the user. The objective shows some beautifully colored Newton rings smack in the middle of the lens, a good sign that the elements are properly centered. Measuring just over a meter long from the tip of its dew shield to the racked-in focuser with diagonal in place, the 'scope tips the scales at a reassuring 3 k. That's right up there with the Zeiss Telementor optical tube, which weighs in somewhere near 3.5 k. Mine came with a shark fin finder bracket, a feature not found on all of the other models. It's a one off, but it looks great! For the record, Skylight has also introduced an all-brass version of the f/15 m. Called the Aureus, it sports the same precision optics in a magnificent (but significantly heavier) tube (Fig. 6.27).

The focuser is a Crayford by design and, unlike many old school 60 mm refractors, can accommodate 1.25-in. eyepieces. Movement is exceptionally smooth, making precise focusing child's play. It also has an adjustable tension knob (the only piece of plastic on the entire scope that could be found), which can come in handy for photographic applications. And with 4 in. of back focus, all eyepieces will perform well. The smoothness of this unit is a far cry from the overly stiff focusers found on the majority of classic 60 mm refractors (Fig. 6.28).

The interior of the tube is painted matte black and is also flocked to suppress any unwanted glare during terrestrial or astronomical applications. The tube is made from strong but lightweight aluminum, with a beautifully finished powder coat for extra durability. Although the dew shield on some Skylight f/15 ms is all black, others are adorned with an all-brass version, such as those supplied with Richard's larger 4-in. instruments.

Fig. 6.27 The appealing Skylight f/15m (Image by the author)

Make no mistake about it – this is a totally redesigned 60 mm 'scope. For example, you might be somewhat surprised to discover that the optical tube assembly is considerably larger than a 'regular' 60 mm refractor. Indeed, with an outer tube diameter of 71 mm, you might have to settle for a makeshift mounting bracket prior to ordering some custom-made rings for the telescope. Richard intends to include these rings (an optional extra now) with all future instruments. That said, you should be able to manage getting a fairly snug fitting with the clamshell for your 80 mm 'scope by padding it out a little. From there, it is a simple matter to mount it securely onto an ashwood alt-az tripod (Fig. 6.29).

Having a cell that can be adjusted by the user is an especially noteworthy aspect of this telescope. Some of the vintage 60 mm refractors from the 1950s and 1960s had such a provision. Alas, that kind of luxury is now as rare as hens' teeth. Even celebrated telescopes such as the aforementioned Zeiss cannot be collimated by the user. With such a small objective, why bother?

Well, based on dozens of examples of the classic 60 mm refractors available out there, it is clear that miscollimation can actually account for the small variation in star tests between classic 60 mm scopes. And almost all of the latter come with optics housed in non-adjustable cells. Experience shows that tweaking collimation – even on these long focus doublets – can turn a good 'scope into one that is subtly better. Indeed, this renewed sense of empowerment, i.e., the ability

Fig. 6.28 Modern touches – the silky smooth Crayford of the Skylight f/15m (Image by the author)

to self-collimate, creates a whole new level of user pleasure. Amidst the many barbarous innovations of our age, it is comforting to divine pleasure from such a simple and well-executed instrument. Hopefully these flower blossoms are not wasted on the desert air!

In the next chapter, we'll be taking a broad survey of the classic 60 mm refractor market and the extraordinary success of the Japanese optical houses that supplied their fine lenses.

Fig. 6.29 Brazen as you like – the Skylight f/15m Aureus (Image credit: Richard Day)

Chapter 7

Classic Specula

Newtonian reflectors remain popular among amateur astronomers because of the unbeatable bang for the buck they offer. Indeed the Dobsonian revolution did much to empower the amateur masses with large, high quality, yet affordable reflectors for the first time in history. And while classic telescopes usually conjure up visions of a spy glass astride an elegant mount, it remains the case that reflectors are highly popular with collectors and restorers. This chapter is dedicated to describing some beautifully made Newtonians from yesteryear and how, after decades and even centuries, they still deliver delightfully crisp images at the eyepiece.

As we have seen from earlier chapters in the book, the British have enjoyed a long and illustrious past as builders of top class refractors. But throughout the Victorian era, Britain was also a world leader in producing some of the highest quality Newtonian reflectors the world has ever seen.

The basic design of the Newtonian reflector – so named because of its invention by Isaac Newton – has hardly changed since it was first conceived by the great scientist in 1668. Instead of using a convex lens to focus light, he used a finely polished spherical mirror. Astronomers had known about the possibilities of parabolic mirrors since 1663, when James Gregory (1638–1675), an English mathematician and astronomer, envisioned a reflecting telescope that would bounce light between two mirrors, one with a hole in it to allow light to reach the eyepiece. Indeed, the Gregorian design pre-dates the first practical reflecting telescope, built by Newton in 1668. Of course, being one of Europe's finest mathematicians, Newton was well aware of the properties of parabolic mirrors that would in theory produce even better images. But methods to 'carve out' a parabolic surface presented a practical problem beyond him at the time. That's why he settled on the less-than-perfect spherical geometry for his metal mirror. The reflected light was sent back up the tube to a tiny flat mirror, mounted centrally and at a 45° angle to the incident rays, delivering the light cone to the eyepiece where it reached focus.

N. English, *Classic Telescopes: A Guide to Collecting, Restoring, and Using Telescopes of Yesteryear*, Patrick Moore's Practical Astronomy Series, DOI 10.1007/978-1-4614-4424-4_7, © Springer Science+Business Media New York 2013

Fig. 7.1 A replica of Newton's reflecting telescope (Image credit: Pulsar Optical)

Newton was apparently very fond of pointing out that his little telescope – which delivered a power of about 40× – performed as well as the non achromatic refracting (lens-based) telescopes many times longer (Fig. 7.1).

Spherical mirrors are easier to make, but they have one minor flaw – light from the edges of a spherical mirror do not come to focus at the same point as rays from the center. In other words, the spherical mirror exhibits *spherical aberration,* which smears out the image so that it is difficult to get a razor sharp view. That said, you can still obtain good results with spherical mirrors so long as the focal length of the 'scope satisfies the following formula:

$$\text{Focal length} = 4.46 \times \left(\text{Aperture in inches}\right)^{4/3}$$

This formula gives the minimum focal length a spherical mirror needs to be in order to meet the *Rayleigh criterion,* which is the lowest quality level that will produce an acceptably sharp image. For example, if you construct a 6-in. (15 cm) spherical mirror, it would need to have a minimum focal length of $4.46 \times (6)^{4/3}$. Plugging these numbers into a calculator gives a value of 48.6 in. (1,245 mm). There are commercially available Dobsonians that have spherical mirrors, but these are usually confined to apertures less than 6 in. for practical reasons.

When you take a mirror that has a nice spherical shape and deepen its curvature at the center a little bit, you will eventually arrive at a parabolic shape. It can be proven mathematically that only a parabolic surface has the attractive property of bringing to a single focus all rays parallel to its axis. In other words, a perfect parabolic mirror would have no spherical aberration. The English mathematician and inventor John Hadley (1682–1744), together with his two brothers, George and Henry, built the first reflector with a parabolic mirror; a 6-in. (15 cm) instrument of 62-in. focal length, which he presented to the Royal Society in 1721.

Although the prize for making the first parabolic mirror went to an Englishman, it was the Scots-born James Short (1710–1768) who would become one of Britain's choicest instrument makers in the mid-eighteenth century. A gifted preacher and theologian, Short's optical skills were also in much demand. His compatriot, the great mathematician Colin Maclaurin, once wrote of his artistry:

> *Mr. Short, an ingenious person well versed in the theory and practice of making telescopes, has improved the reflecting ones so much, that I am fully satisfied he has far outdone what has yet been executed in this kind. He has not only succeeded in giving so true a figure to his speculums of glass quick-silvered behind, as to make the image from them perfectly distinct, but has made telescopes with metal speculums which far surpass those I have seen of any other workman.*

It pays to remember that in those days, telescopes with high quality optics were somewhat of a rarity, especially in large sizes. James Short was a key figure in developing ways of increasing the size of parabolic mirrors. The majority of his telescopes were Gregorians, which employ a concave parabolic primary mirror – and conveniently produce an upright image – and ranged in size from small, hand-held 'perspectives' to instruments as large as 18 in. fitted inside tubes up to 12 ft long. We witness in Short's work the beginnings of what might be called 'precision technology.' He was, for example, the first (after Hadley) to produce consistent paraboloids to correct for the spherical aberration that plagued the high power views of the more common spherical mirrors of his day. What's more, Short was rather secretive about the methods he employed to figure his mirrors. Indeed, he is said to have destroyed all his tools before his death – an act of professional jealousy perhaps? (Fig. 7.2).

Short's smaller optics almost invariably had elegant brass tubes and were mounted on simple tabletop alt-azimuth stands. With each instrument came a terrestrial eyepiece, together with several astronomical oculars. Because Short's instruments often exceeded the quality of those produced by other opticians, he could command higher prices for his works. Indeed, he was so successful that he is reputed to have left £20,000 – a vast sum at that time – in his will. Such recognition led to his election to the Royal Society in 1737. Indeed, Short produced a number of instruments for the society and communicated the observations he made with them. Several of his telescopes and other instruments were dispatched by the society to observe the 1769 transit of Venus, while Short himself observed the transit, surrounded by members of the British nobility, at Savile House, England (Fig. 7.3).

Other makers that are highly sought after by collectors of antique telescopes include George Adams (1734–1772), who in 1760 rose, by the tender age of just 26, from obscurity to become Instrument Maker to His Majesty King George III. Setting up a workshop in London, the exceptionally talented Adams produced telescopes that are true art forms. After his untimely death at age 38, he was succeeded

Fig. 7.2 A Gregorian telescope of c. 1735 vintage (Image credit: Sage Ross)

Fig. 7.3 A 4-in. reflector with mount and eyepieces made by George Adams, Sr (Image credit: Fleaglass.com)

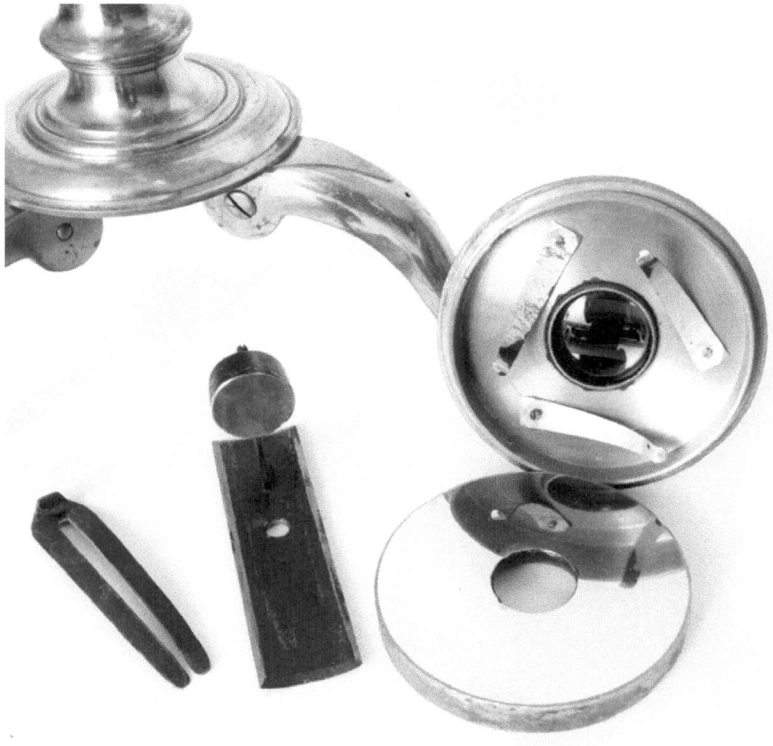

Fig. 7.4 The innards of a Gregorian telescope by George Adams, Sr (Image credit: Fleaglass.com)

by his son George Adams, Jr. (1750–1795), who, as well as carrying on the telescope-making business, would also publish influential works on microscopy, geography, astronomy and physics.

When George Jr. died, his brother Dudley acquired all the stock at auction, thus maintaining the trade name until 1830. Although at first all the instruments of George Adams, Sr., carried the manufacturing date, after 1760 only the name *G. Adams London* appeared, rendering precise dating a near impossibility (Fig. 7.4).

For nearly two centuries after the invention of the Newtonian, the mirrors were made from a special alloy of mainly copper and tin. These 'speculum' mirrors (nearly 62% copper and 38% tin) gave a golden cast to the image and had a reflectivity of about 70% (actually, a 1947 study suggested that its reflectivity varied from about 63% for blue light to 75% for red). After 6 months of exposure in a damp climate, its reflectivity drops by 10%, necessitating frequent polishing. Coupled to this, metal mirrors are exceedingly difficult to grind and are unduly heavy for their size. These deleterious aspects of speculum mirrors forced astronomers to look for better ways to build mirrors (Fig. 7.5).

Fig. 7.5 Proto-Dobsonians? Reflectors evolved into a great variety of forms throughout the nineteenth century (Image credit: Robert Katz)

From Speculum to Glass

At the Great Exhibition of 1851 a number of curious items were displayed – globes and vases silvered on the inside in a process that had just been patented by Messrs Varnish & Mellish. The vessels had been filled with a solution of silver nitrate to which grape juice was added. The fructose present in the solution slowly reduced the silver ions, transforming them into tiny particles of silver that were deposited on all the surfaces in contact with the fluid.

When news of the process reached the ears of the German chemist Justus von Liebig (1803–1873), he immediately understood its importance and potential, and proceeded to modify the process to increase its efficiency. But it was not until 1856 that the German astronomer, Karl Steinheil (1801–1870), used a similar procedure to coat a 4-in. (10 cm) glass mirror with a thin veneer of silver. The telescope, by all accounts, gave excellent images. The following year, the physicist Jean Foucault made his own silver-on-glass mirror, and the resulting telescope – together with the tests he singlehandedly developed to test its quality – received unanimous praise from the French Academy of Sciences.

The basic principle of depositing silver on glass actually forms the basis of a test for reducing sugars. Indeed this author has demonstrated the basic technique many times to students. Silver ions (Ag^+) react with the hydroxide ions (OH) formed in aqueous ammonia to produce a brown precipitate of silver oxide (Ag_2O), which is dissolved by adding an excess of aqueous ammonia. This results in the formation of silver diamine $[Ag(NH_3)_2]^+$. In the final step, the diamine is reduced by glucose to form metallic silver. Glucose is oxidized to gluconic acid (Fig. 7.6).

If this reaction (known collectively as Tollen's reagent) is set up on a clean, smoothly polished glass surface, it results in a layer of silver being deposited on the surface (Fig. 7.7).

These innovations set the scene for the rapid elevation of the reflecting telescopes in both the amateur and professional astronomy circuits that have continued unabated to this day. Plate glass mirrors could be made lighter and so more easily mounted inside their tubes. What's more, because parabolic mirrors work

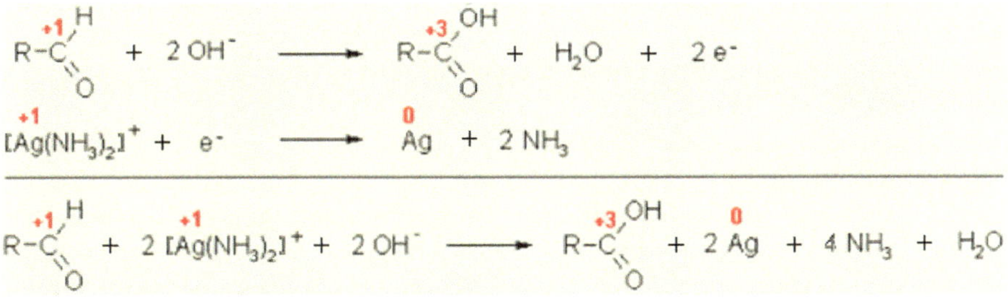

Fig. 7.6 The reactions involved in the silver mirror test for a reducing sugar

Fig. 7.7 A thin deposit of silver forms when a reducing agent is added to Tollen's reagent

well, even at short focal ratios, they could be made much smaller than the standard instruments of the day – the long focus classical achromatic refractor – and thus were often more convenient to use in the field.

The new technical advances in mirror making soon crossed the Channel and were enthusiastically endorsed by instrument makers of note. The retired Hereford schoolmaster and amateur optician George Henry With (1827–1904) began making silvered-glass mirrors as early 1860s and sold them to customers, the most notable of which was John Browning (1835–1925), a London instrument maker celebrated for the spectroscopes and telescopes that left his workshops. Browning was a contemporary of another immortal figure in mid-Victorian telescope making, George Calver (1834–1927).

Calver was born at Walpole, a tiny hamlet in the English county of Suffolk. Raised the son of farm laborers, he became an orphan while still a lad. But despite being condemned to a life of abject poverty throughout his youth, his skill was recognized early, and by the 1850s he secured an apprenticeship to a local shoemaker. Shortly thereafter he set himself up in business at Great Yarmouth, where he met his future wife, Hannah.

It is uncertain as to whether Calver had developed an interest in astronomy in his youth, or whether he had even looked through a telescope, but we do know that the course of his life changed utterly and forever by a meeting with a one Rev Matthews, a local non-conformist clergyman, who allowed him to look at some of the showpieces of the sky through his excellent reflecting telescope powered by a high quality With mirror. Calver wondered whether he could learn how to make mirrors as good or better than With, and when challenged by Matthews, Calver embarked upon an occupation that would both consume his intellect and sustain him financially for the remainder of his life.

Calver absorbed the best literature of the new mirror-making techniques of the day. He also began a close and regular correspondence with G. H. With. Remarkably, although both had become rivals in business, they maintained a healthy respect for each other and indeed shared key ideas.

Calver's first telescope was a 10-in. Newtonian, which he kept for his own use to carry out extensive studies of Jupiter and double stars. But he soon began to receive commissions to build other mirrors in the range of 5- to 8-in. diameter, and in relative apertures of f/9 to f/12. Calver produced Newtonian and Cassegrain configurations to meet the contemporary demand in the market. By 1871 his mirrors were highly sought after, necessitating both a move to new and larger premises at Widford, on the outskirts of Chelmsford, as well as the hiring of a small staff. Although he employed machinery for grinding and polishing his mirrors, Calver always completed the figuring work by hand. What's more, he rigorously tested his mirrors by pinhole, knife edge and eyepiece methods. In like fashion to With, he also constructed a highly polished black glass ball placed a few hundred yards away to produce a point-like source, or artificial star, for optical evaluation. All final testing ended with the stars, using high-powered eyepieces.

Being an experienced observer Calver was only too aware of the thermal issue mirrors suffer as they acclimate to the outside air, and, accordingly, he introduced steps to ameliorate them as far as possible. For instance, most of his mirrors were slightly under-corrected to compensate for the natural over-correction that occurs during night-time cooling.

By the 1880s, Calver's mirrors were in huge demand by a multitude of both amateurs and professionals across the British Empire. And although most of his orders were for smaller apertures of 10–15 in., he also secured commissions to build mirrors up to 37 in., the most famous of which is still in use today, installed on the Crossley Reflector (1895) at the Lick Observatory in Chicago.

Calver was so preoccupied with making his prestigious mirrors that he had little time to devote to their mounting. Thus, he subcontracted those duties to Messrs T. Lepard & Sons, a firm he became familiar with in his days at Great Yarmouth.

By the mid-1880s, Calver was commissioned by Sir Henry Bessemer to make a 50-in. diameter mirror to his own, cost-cutting specifications. The project was a disaster by all accounts, and it had been suggested that the great mirror be broken up to make a number of smaller mirrors. Shortly before his death in 1876, Mr. James Lick, the super-rich American patron of astronomical research, offered a prize for a world record-sized mirror. Calver caused a sensation in the telescope building world by offering to produce a 100-in. behemoth mirror. Unfortunately, Lick took him up on the offer. Perhaps the American tycoon never really believed it was technically possible to deliver such an enormous optic, or perhaps he doubted Calver's ability. We'll never know for sure, of course, but it is worth bearing in mind that men of ingenuity were thinking of constructing such huge telescopes some three decades before G. W. Richey and his team finally unveiled their 100-in. mirror in 1917 for the Hooker reflector atop Mount Wilson Observatory in California.

Fig. 7.8 A refurbished Calver reflector on an alt-az mount (Image credit: Robert Katz)

The sheer rate at which the ugly head of urbanization was unfolding across England at the turn of the nineteenth century forced Calver's hand to relocate to the rural tranquility of his native Walpole. In 1904, he purchased a large house (The Manse) in the village where he continued to churn out top-quality mirrors, although on a much reduced scale. Indeed, Calver continued working into his 90s and died on July 4, 1927, followed by his wife just a year later.

John Calver will be fondly remembered as one of England's finest opticians, producing over 4,000 mirrors in his long career. Indeed, Calver's telescopes are still seeing starlight today in the hands of amateur astronomers who have lovingly restored them to full health. Robert Katz, an amateur astronomer based in London, England, was kind enough to show this author his wonderful 10-in F/8 Calver 'Dob,' which he presses into service from his back garden when conditions allow (Fig. 7.8).

My 10-in F/8 Calver reflector looks like an unwieldy beast, and by any modern standards it is overwhelmingly long. The original wooden stand had rotted and was missing its slow motion controls when I found it, but luckily Len Clucas, the former professional telescope-maker for Grubb Parsons in Newcastle, had inherited an identical stand and cradle from the late master mirror maker David Sinden, which he refurbished for me. A stepladder is essential for objects over 30 degrees high and viewing near the zenith is positively dangerous. And yet – climbing up to the eyepiece apart – it is remarkably easy to use. The eyepiece is always in a convenient position – assuming you can reach it – the azimuth and altitude controls are smooth and make tracking easy even at powers of 300x and the ingenious system of a clamped tangent arm makes rewinding the azimuth screw simple without losing position.

Even though it weighs a ton the telescope is also beautifully balanced; unclamped from the slow motions, with a 40 mm eyepiece in the barrel, I imagine it is the closest you can get to the laid-back star-hopping Dobsonian experience with Victorian equipment.

The optics are fine, and because the focal length is actually less than that of a standard SCT, views of deep sky objects are impressive with a low power eyepiece. It comes into its own with the planets, though, and the exceptional opposition night of Jupiter in September 2010 was memorable in many ways. Thanks to good seeing in southwest London – the telescope is in Hampton Hill – I spent most of the night watching Jupiter turn in exquisite detail using a fine telescope made in 1882 by one of the two great telescope makers of his day. But a telescope so simple that a child can learn to operate it confidently in five minutes.

American Rivals to British Supremacy

The best silvering processes, while effective, still required heating of the glass surface with the real possibility that it might shatter the glass. As we saw previously, John A. Brashear was a gifted telescope maker, employing the latest methods to grind, figure and polish glass. In his autobiography, *A Man Who Loved the Stars,* Brashear describes, in some detail, the construction of a 12-in. mirror of 10 ft focus. But after adding the boiling hot silver solution and reducing agent, he watched in horror as the mirror cracked from edge to center! This experience impelled him to seek a better way of depositing silver on his finely made glass mirrors.

Starting out with only a rudimentary knowledge of chemistry, Brashear, through hard work and lots of experimentation, managed to greatly simplify the silvering process, while also improving its efficiency. Brashear found a way to deposit the silver at the same temperature as the mirror surface (normally about 18 C), thereby avoiding the use of high temperatures. After completing a second 12-in. mirror he wrote:

I had made quite a number of experiments with varying methods of silvering by this time, and at last I found a method, or rather a modification of a method, which I had seen in Scientific American, *called Burton's method, by which I succeeded in obtaining most admirable results in silvering mirrors on the front surface, although it was originally*

Fig. 7.9 Built like a tank, a Brashear reflector survives the ravages of time (Image credit: Dan Schechter)

intended for back surfaces, looking-glasses etc. So simple, and so certain was this method that I at once sent a communication to the 'English Mechanic and World of Science,' describing it in full for the benefit of my amateur friends, of whom there were at that time, literally speaking, scores who were trying to make their own reflecting telescopes. Little did I think at the time that this method would become THE method, and be universally used for front surface mirrors.

The Brashear Process, as it came to be affectionately known, was not superseded until the 1920s, when mirror makers switched to the more efficient, vacuum deposition process.

Dan Schechter, whom we met earlier in connection with classic American refractors, was kind enough to share some images of his 8-in. Brashear Newtonian. Typical of the day and age, these reflectors were built like proverbial tanks, molded from steel and brass. But this author was curious about its performance after all these years. And there's an interesting story that goes with it (Figs. 7.9, 7.10, 7.11, 7.12, and 7.13).

Fig. 7.10 The all-important Brashear logo (Image credit: Dan Schechter)

Fig. 7.11 The sturdy but supremely functional focuser on the Brashear reflector (Image credit: Dan Schechter)

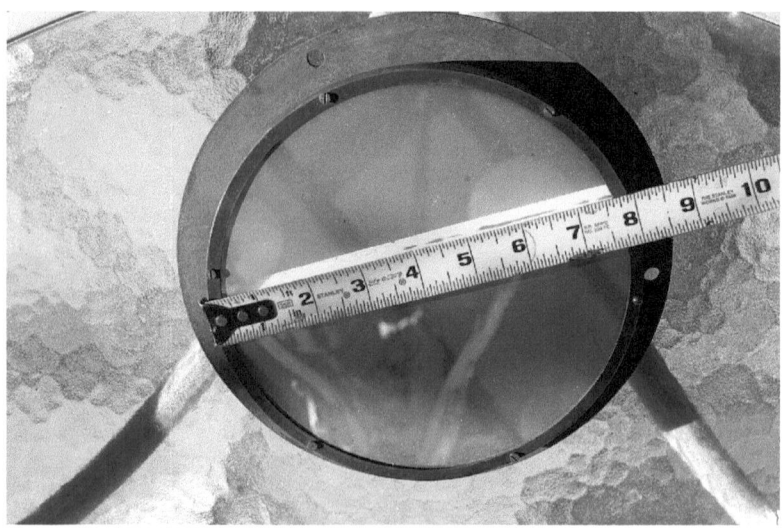

Fig. 7.12 The finely figured mirror on the Brashear 8-in. Newtonian (Image credit: Dan Schechter)

Fig. 7.13 A close-up of the beautifully crafted housing for the Brashear secondary mirror (Image credit: Dan Schechter)

The first time my friend Clint Whitman and I collimated the 'scope, we both could not wait for the Sun to set. We aimed the telescope at a bright star and discovered that the only eyepiece that would come to focus was a 25 mm one. Any other shorter focal length eyepieces would not come to focus. I would get to "best" focus and would go past it and then come back without ever getting a decent star image. The in and out star patterns looked like kidney beans. To say the least, I was devastated. I came to the conclusion that some prior owner had tried to "improve" the figure and totally messed it up. I took the mirror to an amateur who lived about an hour from me for testing and as expected it tested horribly while in its metal cell. We spent most of our allotted time setting up the test and doing the test in the cell and then analyzing the results with a computer program. With very little time left for additional testing, I took the mirror out of the cell and it looked much improved, but we did not have any time to run the results thru the computer program. However, the visual appearance with a Rhonchi grating looked optimistic. I then took the mirror to a friend who works at the Mt. Wilson Observatory and we stripped off the coatings and I left it to be aluminized in the next run. He phoned me a couple of weeks later and I picked it up.

I then took the mirror to Clint's house and placed it in the cell. However, I removed a pie tin that had supported the mirror. My guess was the pie tin took up too much room and the mirror got pinched when I screwed in the screws that held the retaining ring in place. Clint and I collimated the optics once more and again anxiously waited for dark. The first object I observed was Saturn. I used the 25 mm eyepiece first and as expected Saturn came into focus. It also came into sharp focus with a 16 mm eyepiece, followed by a 10 mm eyepiece followed by a 7 mm eyepiece which yielded a power of about 230×. All came to sharp focus and I could easily observe several of Saturn's moons. To say the least, I was ecstatic!

Dan has since sourced a mount for his Brashear reflector that is consonant with the period. As you might expect, such instruments necessitated a massively built mount to stabilize the views during astronomical use. After full restoration, it should be magnificent to look at and through! (Figs. 7.14 and 7.15).

The Incomparable Cave

The traditions of fine American mirror making continued unabated into the twentieth century. One notable optician deserving mention is the late Tom Cave (1923–2003). After experiencing the horrors of World War II in Europe, Cave returned to his home in Long Beach, California, and enrolled in the University of Southern California, where he pursued the study of optical engineering.

Cave designed his first telescope at the tender age of 14 to study the planets, and while still an undergraduate, he made a string of high quality mirrors in the 6- to 10-in. aperture range for amateurs in Los Angeles. Indeed, according to one source, Cave had made more than 900 mirrors by the time he was 37!

It was in December 1950, when he was just six credits short of his degree, that Cave decided to leave full time education to establish, along with his father, Cave Optical Company, run from his garage in Long Beach. Shortly thereafter, however,

Fig. 7.14 A massive equatorial mount typical for a Newtonian of the late nineteenth century (Image credit: Dan Schechter)

space constraints at home forced him to rent a larger premises, situated just a mile down the road. Cave, Sr., was a skilled mechanic and built several modified Draper polishing machines for his son's burgeoning business. At first, Cave Optical only produced primary mirrors to order, but as their enterprise grew, the company also added tube assemblies and the mounting systems to their inventory, and by the 1970s Cave had 30–35 employees, including eight opticians (Fig. 7.16).

Cave pioneered the mass production of high quality telescopes, and his reputation led to the production of large, observatory-class telescopes and government contracts. For example, Cave was commissioned to produce several mirrors for NASA, the largest having a diameter of 30 in.. Nor did Cave confine his efforts to Newtonian optics. The company also produced Cassegrain telescopes of various sizes, and they even figured and polished primary mirrors for Questar Maksutovs

Fig. 7.15 Schechter's period-correct mount to be mated to his Brashear reflector (Image credit: Dan Schechter)

(discussed in a later chapter). Indeed, over a period of 30 years, Cave Optical would produce over 83,000 mirrors and 15,000 complete telescopes!

The rise in popularity of the compact and economically priced Schmidt Cassegrain telescope, as well as more aggressive competition from other telescope manufacturers, conspired to bring Cave's business to an end in 1980. That said, Cave's instruments are still celebrated for the quality of their smaller 'scopes, especially the Cave Astrolas, which are still used and highly valued among planetary observers today.

Another U. S.-based company, Edmund Scientific, has enjoyed a loyal following from Newtonian fans ever since its inception back in 1942. Founded by the late

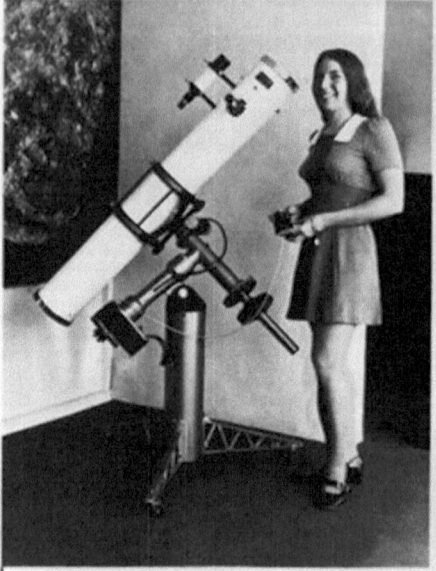
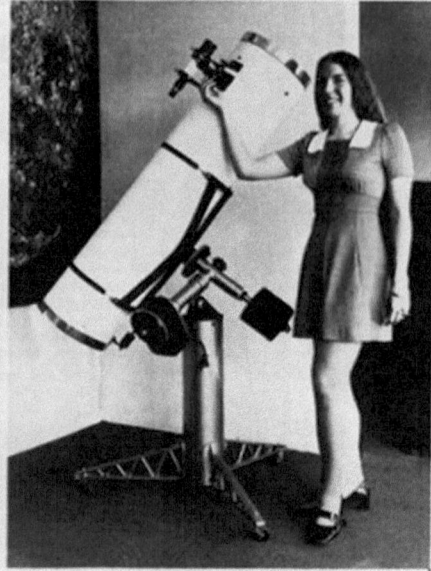

Fig. 7.16 The highly successful line of Cave Astrola Newtonians

Fig. 7.17 A c. 1958 vintage Edmund 4.25-in. f/10 Palomar (Image credit: Al Paslow)

Norman W. Edmund (1916–2012), the firm was originally called Edmund Salvage Co., based in Barrington, New Jersey, and originally sold surplus optics. This was quickly followed by the offering of a number of complete telescope models for sale which, by the 1960s, had diversified its line into several popular models. These included the Space Conqueror series, which included a 4.25 in. f/10 Newtonian reflector with a spherical mirror (Fig. 7.17).

The company initially offered a 6-in. f/6, and in the 1970s also produced a slower, 6-in. f/8 instrument with a parabolic mirror and a substantial equatorial mount. One enthusiastic owner of both these instruments stated, "These are great 'scopes with very fine optics. Early Edmunds instruments have remarkable mirrors for the money, although they are rare beasts today."

The company also sold a larger 8-in. f/8 reflector complete with a 1.25-in. focuser, 6×50 finder and heavy duty equatorial mount. A clock drive could also be purchased as an optional extra. Perhaps the most fetching of all of Edmund's larger instruments, though, is the Edmund Model 4001, an 8-in. F/5 Newtonian on a newly designed fork mount, complete with a smooth, 2-in. focuser to accommodate wide-angle eyepieces that were then available (Fig. 7.18).

Few amateurs of long standing will be unfamiliar with Edmund Scientific's little Astroscan, or Model 2001, as it was otherwise known. Launched in 1977, it brought an award-winning 4¼-in. f/4 rich field Newtonian to the masses. A red plastic shell and optical window encased a very portable richest field telescope with a 3½ degree field of view. The instrument sat on a little ball and socket mount that could be

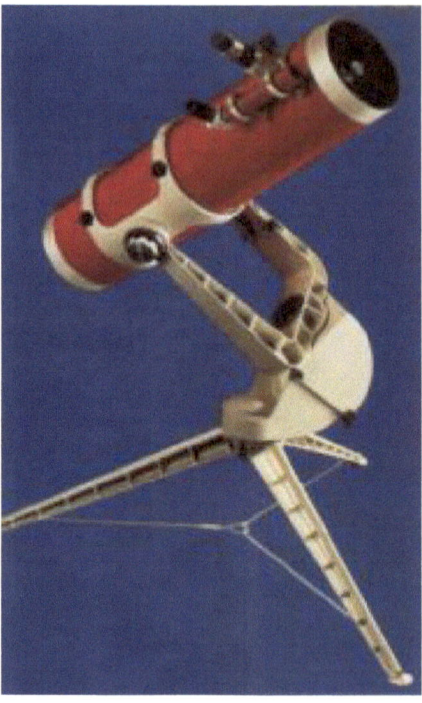

Fig. 7.18 The Edmund 4001 (Image credit: Edmund Scientific)

quickly positioned to point at any object in the sky. It was adored by many but derided by others, who claimed it was difficult to aim, lacked the ability to be collimated by the user and rendered images that were fuzzy at high powers.

Edmund Scientific recently re-launched their Astroscan, in a supposedly improved model called, rather unimaginatively, the Astroscan Plus. At its heart is a 4.1-in. F/4.3 parabolic mirror. What's more, the website states that this mirror has a figure of 1/8 wave and so theoretically should be capable of pretty decent low and high power images, despite its 36% central obstruction.

Astroscan Plus comes with two upgraded eyepieces – a 28 mm Plossl yielding 15× and a 15 mm Plossl serving up a power of 30×. These are a noticeable improvement on the original Kellner eyepieces that came with the earlier versions of the 'scope. The focuser has also been fitted with Teflon bearings to improve its movement and is also supplied with a unit power red dot finder. Weighing a mere 13 lb including the base, the instrument can literally be taken anywhere at a moment's notice. For wide-field vistas in a hurry, this 'scope is hard to beat (Fig. 7.19).

A review of the new and improved Astroscan Plus conducted by Gary Seronik in the July 2010 issue of *Sky & Telescope* magazine revealed that the 'scope worked well for low power, wide-field sweeping, but images became 'noticeably soft' at powers above 50×. He noted that the most likely cause of this compromised performance at

Fig. 7.19 The newly upgraded Edmund Scientific Astroscan Plus (Image credit: Edmund Scientific)

high powers was a slightly miscollimated optic. That said, the Edmund Scientific Astroscan Plus still comes with a 5-year warranty (like the original), so any performance issues, should they arise, can be resolved quickly. The Astroscan Plus can be purchased as a standard ($229 plus shipping) or deluxe package ($429 plus shipping). The standard package gives you everything described above, as well as a slip-fit dew cap, an up-to-date 36-page User's Guide, an adjustable shoulder strap, a 35-page "Sky Guide" booklet and a star & planet locator. The deluxe package also includes a roof prism, which orients objects right-side up and left-to-right, a Sun-viewing screen, a 2.5× Barlow lens, a nylon tote bag and a lens cleaning kit.

The Criterion Dynascopes

Edmund's main competitor in the 1960s was the Criterion Manufacturing Company of Hartford, Connecticut, and the impressive line of medium aperture Newtonians they produced. Those of us old enough to remember will recall the full-page ads Criterion had in every issue of *Sky & Telescope* magazine. The company actually began a decade earlier, when they offered a small (sub-3-in.) reflector for sale. After establishing a viable market, the company soon progressed to the 4-in. Dynascope,

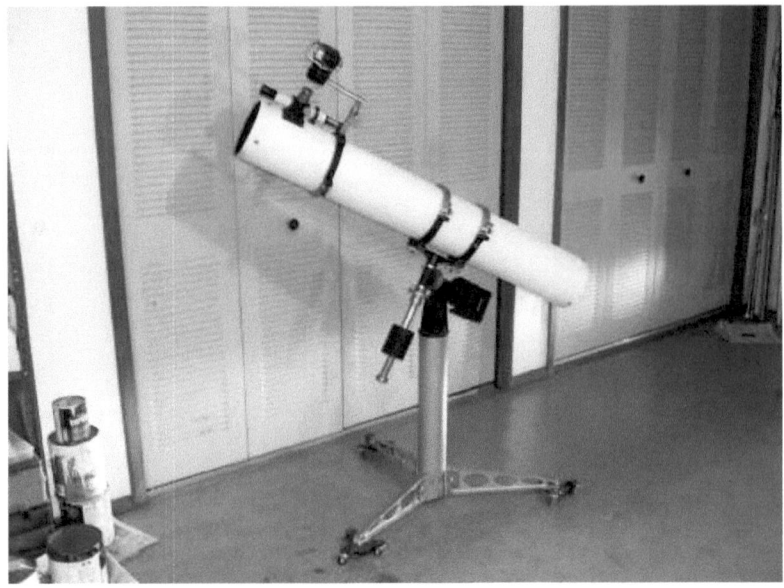

Fig. 7.20 In good working order; the old RV-6 still has the power to turn heads (Image credit: Bill Nielsen)

and from there to a suite of beautifully crafted Newtonian reflectors on heavy duty German equatorial mounts and in apertures ranging from 6 to 12 in.. A 16-in. instrument was allegedly produced by the company, but this author has yet to hear of one in actual use.

Make no mistake about it, these were very expensive telescopes when they were first launched in the late 1950s. Even the smallest instrument in the company's product line, a 6-in. f/8 Dynascope, went on sale for a whopping $475, which, to put it in perspective, was nearly seven times the average weekly wage for an American worker! (Fig. 7.20).

Wishing to boost sales of its 6-in. Dynascope, Criterion decided to offer a stripped-down model just a year after launching the product. This 'no frills' instrument came without the pier mounting, drives and setting circles and was offered at a reduced price of $265. The strategy was not entirely successful, though, especially since other manufacturers were churning out high-quality instruments at prices that were significantly lower than even their stripped down Dynascope. The response was ingenious and manifested itself in the form of the venerable RV-6, which made its debut in June of 1959.

Compared to the original 6-in. Dynascope, the RV-6's mount was less robust but was driven by a quality electric clock drive. The tube end-rings had disappeared, and significantly, had no permanent pier but were instead supported by a pedestal

Fig. 7.21 Bill in the 1960s, projecting an image of the Sun with his RV-6 (Image credit: Bill Nielsen)

similar to pedestals sold by Edmund. Other luxuries also disappeared, including the 50-mm finder. But for the eye-catching price of $194 you got a terrific deal – a sturdy German equatorial mount with rotating tube, setting circles, three eyepieces, and a 6×30 finder.

One amateur astronomer described the RV-6 as the '57 Chevy of telescopes! (Fig. 7.21). Another, Bill Nielsen, a retired U. S. Coast Guard pilot who now lives in the Tampa, Florida, area, shared his experiences of this telescope:

The 6-inch f/8 optics on the RV-6 provide excellent color free, contrasty planetary images and beautiful clean star images. The 'scope is lightweight and only takes a few minutes to set up. However it's not perfect mainly because it was offered at such a low cost. Mechanically the mount is adequate; the drive tracks well but there is lots of play in the worm gear that makes it difficult to center an object. With practice, it works fine. The focuser and draw tube is another weak feature. Depending on the eyepiece, the draw tube must be depressed or extended to obtain best focus, and is a less than friendly design. The rack and pinion is not precise and takes experience to obtain sharp focus. It takes some skill and practice to use high power and obtain focus while keeping the target in the field of view. As a teenager, I perfected those skills but can understand why someone today would think this very humble at best.

I did a side by side comparison viewing Jupiter with a very expensive semi APO 6-inch f/8 refractor and the RV6 won with better contrast, sharpness and color correction. That comparison led me to sell the refractor and restore my RV-6. Today, although I'm not impressed with its mount, focuser, and finder, I find the optical tube assembly (lightweight tube) and superb optics to be very pleasing. The optical excellence is the RV-6's best attribute.

In this chapter we have explored something of the wonderful world of the Newtonian reflector and the great care and attention folk have devoted to restoring them to full working health. Whether it's a Cave, a Calver or Criterion, these telescopes all serve as reminders of the optical excellence of Newtonian reflectors from yesteryear and why they will be used by amateurs so long as humankind desires to look skyward. In the following chapter, we shall take a look at one of the most celebrated of modern classics, the Unitron refractors of yore.

Chapter 8

The Age of Unitron

Who would have thought that within only a few years of the ending of World II, the defeated Japanese nation would be opening up their optical houses and selling their finest wares to their erstwhile foes in America? Such is the crazy world of international commerce! The early 1950s witnessed a remarkable resurgence in the marketing of a range of small and medium aperture achromatic refractors to meet the needs of a growing army of amateur astronomers across the United States and Canada. Fine Japanese-made achromatic optics, ranging in size from 2 to 6 in. (50–150 mm) found their way across the Pacific, where they were housed in exquisitely made optical tubes and marketed under a number of brand names, including Royal Astro, Mayflower, Sears, and Swift, among others. But it is arguably the 'Unitron' appellation that has become most iconic of this mini-age of refractors (Fig. 8.1).

With a name sounding more like something from an episode of *Dr. Who*, the Unitron brand has gained a well-respected reputation among avid collectors of modern classic telescopes. Dealing almost exclusively with long focus, classical achromatic refractors, these telescopes are some of the most sought after classic instruments out there. And it's easy to see why. Their simple and durable long focal length optics and their rugged and elegant mechanics make them such a joy to look at and use.

Unitron has led the way to the stars for many of the world's astronomers over a certain age. Beginning in the early years of the 1950s the United Trading Company took the designs and the ideas of packaging and accessorizing telescopes from the German company Zeiss and developed it into a series of telescopes that many more amateurs could afford. And like Zeiss and its travel telescopes, Unitron developed refractors that one could keep in a storage box, load it in the trunk of your car and ferry off to great astronomical adventures.

N. English, *Classic Telescopes: A Guide to Collecting, Restoring, and Using Telescopes of Yesteryear*, Patrick Moore's Practical Astronomy Series, DOI 10.1007/978-1-4614-4424-4_8, © Springer Science+Business Media New York 2013

Fig. 8.1 A Canadian postage stamp issued in 2002 honoring Edmond Caillard and featuring a classic Unitron refractor

Over the years, Unitron supplied some very well corrected air-spaced achromatic objectives and housed them inside telescopes varying in diameter from 40 to 150 mm (1.6–6 in.). The smaller 1.6- and 2.4-in. model Nihon Seiko telescopes were provided with one focus knob, accessible from the right side of the focuser. The larger units had the now standard two-knob configuration. In addition to their long, straight-tubed refractors, the company also made folded refractors that were more manageable and so easier to mount, though a question mark remains as to how good they were in comparison to the traditional forms.

Nihon Seiko Kenkyusho, Ltd., of Nozawa, Setagaya-ku, Tokyo, Japan, originally manufactured Unitron refractor tube assemblies for U.S. distribution, of 3-in. apertures and possibly other sizes. Curiously, the Unitron brand was sold under a different name in Europe – Polarex. The objectives for the 4-in. Unitron objectives were produced by many sources. In order to have something to sell that was comparable to the lower-priced imported telescopes being marketed through the 1980s, Unitron ordered refracting telescopes of simpler construction. These telescopes were not made by Nihon Seiko but by other vendors and were more akin to the 'department store' models being sold at the time – a far cry from the values that up to then could be expected of the Unitron name. At least some of these were made for Unitron by Towa Optics of Japan.

Barry Greiner of D & G Optical, Pennsylvania, has examined hundreds of Unitron lenses and their accompanying cells and so has a wealth of first hand information concerning the quality of their optics, as well as how they have changed over the years. The original cell design dating from 1952 to 1954 had just three flat-headed screws, and each non-threaded retaining ring was custom sized to hold the lens in its cell with just the right amount of space so as not to alter or apply pressure to the objective. The first alterations occurred around 1954, when the

Fig. 8.2 A close-up shot of a 3-in. Unitron objective (Image credit: Richard Day)

company added three flat-headed lock screws at the side of each retaining screw. These push out the retaining ring and lock it in place. In addition, they started to hand stamp the Unitron logo on the retaining ring.

The general consensus is that the 1950s were the company's golden decade for optics. They seem to have been hand figured and polished to perfection. That said, during the company's 30 years of supplying telescopes, they were quite consistent in making good optical quality objectives (Fig. 8.2).

Clint Whittmann, a passionate amateur astronomer from California and avid collector of classic telescopes, provided his take on the Unitron experience:

This, as with most American astronomers, was my first encounter with what has now become one of the most collected telescopes in the classic telescope collector's world. We were exposed to these ads as kids, whether it was through the pages of our favorite magazines or at a local telescope store. I remember growing up in the San Fernando Valley. My folks were not rich by any means. I guess we were middle class. To me, this meant we had a 'Tasco' income. So, on my 9th birthday, that was the telescope I received. Granted it did look a little like a Unitron, and I was more than happy with my giant 60 mm 9TE in 1968. I spent countless hours looking for Neil Armstrong walking around the Sea of Tranquillity! This telescope started me out in astronomy. Round about the same time I was able to get my hands on Sky & Telescope *magazines and would spend many wonderful hours looking at all the great telescope ads, but it was the Unitron ads that packed the most punch visually and made me imagine (Fig. 8.3).*

As I look back at these ads, I now recognize the marketing genius that this company was employing. It is not surprising that they sold as many of these great telescopes as they did. Looking through the ads today, one realizes that they actually were pioneers in the credit

Fig. 8.3 The author's fine 80 mm f/11 achromat made by Nihon Seiko, Japan (Image by the author)

and media markets. The telescopes could be purchased by a hire purchase scheme involving monthly down payments. This was the only way many a gentlemen with a job could afford a telescope that cost more than the family car and make a purchase of this size in the 1950's or 60s. The ads promoted and developed amateur astronomy as an outreach, family activity and many of them depict a man, his Unitron, the countryside, a dark sky, the children and maybe a pipe and a dog or Christmas tree. The father was pointing into the sky while junior and his sister are peering into the heavens through the eyepiece of a 4-inch Unitron refractor. What could be better than that for all the parties in the depiction? The sheer size and the look of a Unitron refractor is another factor that makes them some of our favorites. As an adult, I have had the good fortune to be able to own a few of these wonderful telescopes.

The company's alt-azimuth and equatorial mounts have also become synonymous with high quality and to this day remain one of the best built and greatest ever made. The 155 mount with its electric clock drive and easy to reach control rods has made the concept of "imagine being at the controls" into the reality.

Each Unitron telescope came equipped with three or four eyepieces, which usually included orthoscopics, Kelners and a more obscure design known as an 'achromatized symmetrical.' All delivered excellent performance in the long native focal lengths characteristic of Unitron refractors (Fig. 8.4).

The Unitron equatorial mounts – and they came in various sizes to accommodate optical tubes of various sizes – are true works of art. Solidly built and beautifully functional, they can be driven either with an electric clock drive or manually, using slow motion controls.

Fig. 8.4 A marvel of engineering and aesthetic appeal, the Unitron equatorial mount (Image credit: Richard Day)

But how do the Unitrons compare to other highly regarded classics of the same genre? Leonard Marek from California has collected quite a few vintage 'scopes from the age of Unitron and gave me his opinion:

Since I was a teenager back in the late 1960's, I used to subscribe to Sky & Telescope *magazine. I recall wishing to own one of those fine telescopes. It was the styling and classy appearance of the telescope that always struck me as beautiful. I had a Sears 3-inch refractor at the time and thought that the Unitrons were supposed to be much better optically than anything else on the market then.*

Well, now that I have owned a Unitron Model 128 (60 mm) equatorial, a 3 inch Polarex and two Unitron Model 160 and 166 (4-inch) equatorial refractors, I know that my Sears 3-inch Model 4-6339A optics were every bit as good as the Unitron! But the mechanical build of the Unitron was far superior to the Sears. For that time period, those Unitrons were probably most likely the best mechanically as well as optically, but then that too was reflected in the price tag.

There were, as we have seen, many competitors with the venerable Unitron brand. And if truth be told, some were every bit as good. Take the Tasco 20TE, for example, a 108 mm f/15 classical achromat, sold throughout the 1960s and the 1970s by Tasco. Fitted with high quality achromatic objectives supplied by Carton and housed in a tube made by Goto, the 20TE produced superlative views of the night sky (Fig. 8.5).

Back in the day, such a beauty would set you back several hundred dollars, which in today's money wouldn't give you much change out of $5,000. No wonder the 20TE is so eagerly sought after by collectors of classicists (Fig. 8.6).

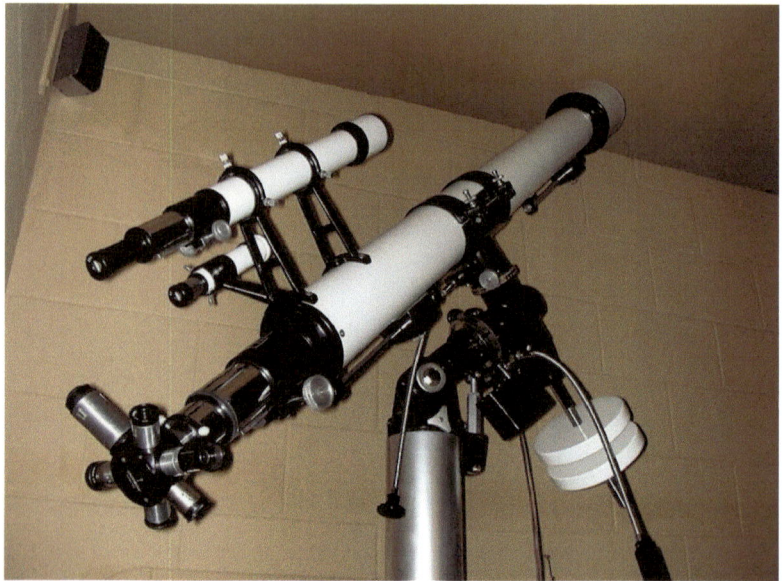

Fig. 8.5 A Unitron simulacrum? The beautiful Tasco 20TE refractor (Image credit: Mike Carman)

End of Days

After the Apollo Moon missions concluded, interest in space exploration and astronomy among the general public declined, with the result that telescope sales, including those of Unitron, were showing signs of drying up. Coupled to that, opticians were experimenting with new glass prescriptions that would offer better color correction and portability.

As we shall see in the next chapter, Takahashi began introducing a new line of super high performance telescopes in the 1970s and early 1980s that began to compete with Unitron. These new Japanese instruments represented the first of a new line of affordable refracting telescopes that were within grasp of the more demanding amateur. Their doublet and triplet objectives incorporated a crown element made of pure, synthetically grown calcium fluorite. This design had much better color correction, and with a focal length some 33% shorter than that of a competing Unitron, the Takahashi refractor was easier to transport and mount. Other companies, such as the American-based Astro-Physics, also began offering apochromats shortly thereafter. Eventually, Unitron was sold to a photographic equipment distributor in 1981, and by 1992 the company was no longer in the telescope business but continued to sell off remaining telescope and accessories inventory for years afterwards.

Fig. 8.6 William Thornton next to his pride and joy, an equatorially mounted 5-in. f/15 Unitron (Image credit: William Thornton)

Chapter 9

Die Zauberflöte

The heart is deceitful above all things, and it is exceedingly corrupt: who can know it?

Jeremiah17:9

Now, what I want is, Facts. Teach these boys and girls nothing but Facts. Facts alone are wanted in life. Plant nothing else, and root out everything else. You can only form the minds of reasoning animals upon Facts; nothing else will ever be of any service to them.

Mr. Gradgrind, from Charles Dickens' *Hard Times*

We have heard much about the milieu of the classical achromat. From the frigid wastes of Russia to the balmy tropical climes of Brazil, the humble crown and flint has distinguished itself as a telescope that can work well in all types of weather. There was once a time where nearly every major observatory across Europe, Asia and the Americas had, at its heart, a large, equatorially mounted refractor engaged in cutting edge astronomical research on the Moon, planets, double stars, not to mention a plethora of dim and distant nebulae. But slowly, the aperture advantages of the reflecting telescope began to supersede the refractor, and as the twentieth century marched on, the role of the achromatic refractor became ever more ancillary, ending its days rather ignobly as part of public outreach programs, or far worse still, having fallen into disuse, or dismantled and auctioned for parts.

Doubtless, the great refractors of yore will never again be built by professional astronomers, but that does not diminish what they have achieved in the past. Even in sizes typically used by amateurs, classical refractors have been largely replaced by shorter focal lengths instruments with improved glass prescriptions. But, as we shall see, the contemporary amateur community lies under a long shadow cast by our telescopic forebears who discovered everything we cherish today using the simpler, crown and flint prescriptions. In this chapter, we shall explore some of the

N. English, *Classic Telescopes: A Guide to Collecting, Restoring, and Using Telescopes of Yesteryear*, Patrick Moore's Practical Astronomy Series, DOI 10.1007/978-1-4614-4424-4_9, © Springer Science+Business Media New York 2013

illustrious deeds of astronomers who used the classical achromat to uncover the unfolding majesty of the universe around us. Thereafter a discussion will follow on some the extraordinary properties of these telescopes, as revealed by new research, with a mind to explaining why they performed so well.

Plumb Line to the Stars

Go outside on an autumn evening and locate the brightest star in Cygnus, the magnificent blue-white Deneb. Now, using ordinary 10×50 binoculars, pan about two binocular fields southeast. Chances are you'll come across a pair of golden suns, separated by a sliver of dark sky. This is the famous 61 Cygni system. A 60 mm refractor provides a splendid view at 38×, the brighter orange star shining with magnitude +5.2 with its fainter, +6.1 companion displaced only 27 arc sec to the southeast.

These form a true binary system with an orbital period of about 700 years. From our cozy vantage point, the pair look relaxed, even serene. But careful inspection of this system over decades and centuries reveals that 61 Cygni is not an ordinary 'fixed star' but is sprinting across the sky.

That much became clear to the Italian astronomer Giuseppe Piazzi, based at Palermo Observatory, Sicily, as early as 1792, when he estimated that 61 Cygni was changing its position relative the background stars by as much as 5.2 arc sec per year. Although that doesn't sound like much – about one eighth of the apparent diameter of Jupiter at opposition – it was enormous by the standards of anything that was observed before. Palermo, however, might as well have been a million miles away from the epicenter of astronomical research in Europe. As a result, Piazzi's observations went largely unnoticed for over a decade until 61 Cygni's extraordinary sojourns were again noticed by the German astronomer Friedrich Bessel, who published a report of the system's large proper motion in 1812. To Bessel, that was a sure sign that these golden suns were relatively nearby, but proving it was quite another matter (Fig. 9.1).

Another astronomer, Wilhelm Struve, director of the Dorpat Observatory in Russia, provided the impetus for Bessel's groundbreaking work. If a star is truly nearby, Struve reasoned, it ought to possess one or more of the following characteristics: it should be fairly bright (nearer stars look brighter); have a large proper motion; and if it happens to be a binary star system, the two components ought to appear widely separated in comparison to the time it takes them to orbit each other. Struve agreed with Bessel that 61 Cygni was an excellent candidate to measure stellar distance. The method to be employed was trigonometric parallax; if a star is close, it should shift its position back and forth against the background stars in the sky, as Earth orbits the Sun.

Bessel was fortunate enough to come of age in an era where astronomical telescopes, especially large classical refractors, were being fashioned to unprecedented standards of accuracy and precision. His compatriot, Joseph von Fraunhofer, used his optical genius to create large achromatic refractors on driven equatorial mounts,

Fig. 9.1 Fr. Giuseppe Piazzi (1746–1826)

an absolute necessity for accurate positional measurements of stars to be undertaken. Thus, in a humble observatory in Konigsberg, Germany, Bessel had installed a purpose-built instrument called a heliometer to measure the parallax of 61 Cygni. Designed by Fraunhofer, it consisted of a 6.5-in. object glass cut down the middle to create two semicircular halves (readers are not recommended to do this at home). The idea actually had its origin back in 1675 with Ole Romer (1644–1710) and was implemented by Dollond in 1754. Each 'half' objective was separately mounted in such a way that one could be moved independently of the other. When perfectly aligned, the two half objectives form a single image, but as one half is moved relative to the other, two separate images are created. The amount of movement needed to superimpose the displaced images can be used to measure the angular separation between two or more objects. Using this method, Bessel measured background stars together with the brighter member of 61 Cygni to deduce the parallax of the star system (Fig. 9.2).

Over a 4-year period beginning in 1834, Bessel subjected 61 Cygni to intense scrutiny, repeating his measurements at least 16 times every night and many more times during nights of exceptional seeing. Conditions could sometimes be cruel, working as he did with his bare hands in an unheated observatory during freezing nights. But he persevered where many others would have given up. His results produced a parallax of 0.3483 arc sec – only 10% less than the modern accepted value – and corresponding to a distance of just over 10 light years.

For the first time in history, someone figured out the immense distance to a star – distances well beyond ordinary human understanding. In recognition of Bessel's

Fig. 9.2 Friedrich W. Bessel (1784–1846)

work, John Herschel, then president of the Royal Astronomical Society, reminded his fellows that they had lived to see the day when the "sounding line in the Universe had a last touched bottom." It was, he continued, "the greatest and most glorious triumph which practical astronomy had ever witnessed."

Establishing the Universality of the Laws of Physics

The observation that distant binary stars orbit their common center of gravity firmly established the idea that the laws of celestial mechanics are indeed universal. This discovery was nothing less than a crowning achievement for Newtonian physics. If you live in the northern hemisphere, cast your gaze on the middle star, Mizar, of the Plough handle of Ursa Major. If you look hard you'll see that Mizar has a fainter 'companion' star – Alcor – set very close to it. But Mizar itself reveals another glorious secret when examined with a modest spyglass; it has a fainter stellar companion tucked up close up to it. Such was made plain to Ricioli in 1650. In 1656, Christiaan Huygens resolved the Trapezium at the heart of the Orion Nebula (M42) yet still did not realize the relationship between its constituent stars. The first person to think of double stars as being gravitationally bound was the English scientist John Michell (1724–93). His ideas were greatly expounded upon by Sir William Herschel, who first thought of them as physical systems and who began research on them a century after the death of Galileo. Indeed Herschel realized that the small changes in position of the stellar companion of Castor and Gamma Virginis were, in fact, not caused by parallax, but by orbital motion about a common center of

Fig. 9.3 Joseph von Fraunhofer (1787–1826)

gravity. In 1802, Herschel first couched the idea of a binary system in a clear and unambiguous way:

If, on the contrary, two stars should really be situated very near each other, and at the same time so far insulated as not to be materially affected by the attractions of neighbouring stars, they will then compose a separate system, and remain united by the bond of their own mutual gravitation towards each other. This should be called a real double star; and any two stars that are thus mutually connected, form the binary sidereal system which we are now to consider.

And while Herschel used his giant speculum mirrors to uncover a great many binary stars, it was his son John (1792–1871), in collaboration with Sir James South, who produced the first catalog of over 3,000 pairs. Yet this impressive bounty of newly harvested celestial treasure was to pale in comparison to the bounteous findings of the Struve dynasty of binary star astronomers, using much smaller telescopes to boot! (Fig. 9.3).

Their success was crucially dependent on advances in telescope optics and mechanics. In 1824, Willhelm Struve (1793–1864) supervised the erection of the finest refractor the world had ever seen. The brain child of Joseph von Fraunhofer, it consisted of a 24 cm aperture achromatic refractor astride a massive, clock-driven equatorial mount and equipped with a filar micrometer. With the great Dorpat refractor, Struve surveyed up to 400 objects an hour. Think about it; in just 9 s, he would center a new object using the finderscope and then examine it at high power before moving onto his next target. In 3 years, having examined 20,000 objects, he discovered a binary system for every 38 stars examined! (Figs. 9.4 and 9.5).

Fig. 9.4 Wilhelm von Struve (1793–1864)

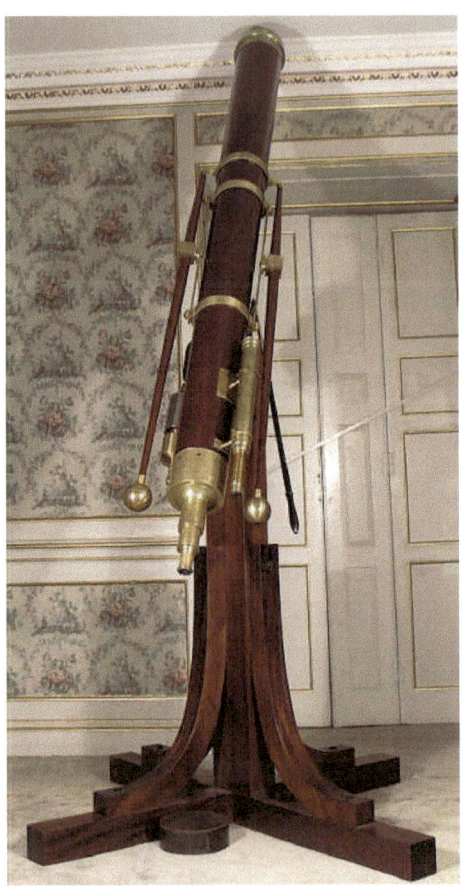

Fig. 9.5 A 7-in. refractor designed by Fraunhofer (Image credit: early technology.com)

In 1839, Willhelm Struve founded Pulkovo Observatory, housing a larger 38 cm refractor, just outside St. Petersburg, Russia, where his son Otto Struve (1819–1905) took on the mantle of his father, discovering some 500 pairs with the great refractor. Indeed, despite nearly two centuries having passed between then and now, roughly a fifth of all binary stars known were cataloged by the Struves.

By the 1870s, new darlings of double star discovery were beginning to blossom in the New World. Prominent among them was Sheldon Wesley Burnham, whom we met earlier in connection with Alvan Clark. With a modest 13 cm refractor, this gifted amateur discovered his first pair in 1873. Once he acquired his 6-in. Clark achromat, Burnham uncovered a further 451 new ones from 1872 to 1877. Burnham's extraordinary success with such a modest instrument embarrassed the astronomical cognoscenti, who had mistakenly believed that essentially all the binary stars visible to the instruments of the day had been discovered. In his entire career, Burnham elucidated no less than 1,300 new double stars.

Other Americans joined in the hunt, including Robert G. Aitken (1864–1951) and William J. Hussey (1862–1926), who conducted their surveys using the two large refractors at the Lick Observatory. Beginning in the autumn of 1899, they began a 5-year study where they discovered a further 2,000 couples. And although they went their separate ways thereafter, Hussey and Aitken continued to catalog and measure the positions of a few thousand more pairs between them.

Aitken's work is particularly noteworthy. You can find numerous entries in the *Publications of the Astronomical Society of the Pacific* from 1900 to circa 1909 of separations of extremely difficult double stars measured using the Lick refractor by Robert Grant Aitken, which have entries ranging from 50 to 100 milli-arcseconds. What's more, these data were used to establish the orbital elements of such binary stars and are broadly accepted today.

Yet, despite its tenfold greater theoretical resolving power and even with the assistance of adaptive/active optics, the Keck telescope atop Mauna Kea can only achieve resolving powers that were, until relatively recently, broadly similar to those achieved by Aitken et al. using the great refractor. How can this be?

Despite the stellar images swimming in a morass of false color, this author suggests that the thermal properties of the glass as well as the slow (f/18) focal ratio of the Lick refractor (the Keck is F/1.75) were the decisive factors in stabilizing the images enough to allow these early-and extremely difficult measurements to be made. Some underlying physics supporting this conclusion will be presented later in this chapter.

Even today, the classical achromat is the instrument of choice for professional binary star astronomers. For instance, the 24-in. Clark refractor at the U.S. Naval Observatory is still used by professional astronomers for binary star work. Yet that did not exhaust the achievements of the classical achromat in the noble art of astronomical mensuration. Beginning in January 1851, George Biddell Airy (1801–1892) used an exceedingly fine 8.1-in. achromatic refractor of 11 ft and 7 in. focal length (f/17), built by Troughton & Simms of London, to establish Greenwich as the Prime Meridian. The same instrument enjoyed continual use until 1954.

The classical achromat also helped humankind divine the constitution of the stars. For centuries, astronomers despaired of ever finding a way to elucidate the chemical makeup of the heavenly bodies. Indeed, the French philosopher Auguste Comte used this very argument in 1844 as an example of a body of knowledge that would forever lie beyond our ken. After all, we could never travel to them and sample them directly. But within 3 years after Comte's death in 1857, his bold conjecture was proven wrong.

Its unraveling actually began in 1802, when the English physicist, William Wollaston, examined a greatly attenuated beam of sunlight through a telescope to which a glass prism had been attached. To his astonishment, he discovered a small number of dark lines in the solar spectrum, but alas, could not offer any explanation for their existence. Within 12 years, Fraunhofer had carefully recorded some 600 lines strewn across the Sun's spectrum. He inched even closer to elucidating their nature when he found identical lines in the reflected light of the Moon and nearby planets.

By the mid-nineteenth century Gustav Kirchoff and Robert Bunsen showed that it was possible to identify a chemical element by matching it to the colors found in the spectrum of the substance under investigation. Soon, it was realized that these colorful displays could provide the key to divining the chemistry of the stars.

Soon, observational astronomers on both sides of the pond were harvesting the first fruits of stellar spectra. In 1863, the English amateur astronomer William Huggins, using an 8-in. Clark objective mounted on a massive equatorial platform built by T. Cooke & Sons, published a paper in the Proceedings of the Royal Society entitled, 'On the lines in the Spectra of Some Fixed Stars.' This was followed by other papers on the spectra of various stars, which showed that each contained a selection of lines – made up of familiar chemical – also visible in the solar spectrum. The distant stars were like the Sun (Fig. 9.6).

In 1842, the Austrian physicist Christian Doppler suggested that the wavelength of light or sound is altered by the motion of the observer or the source either towards or away from each other. Specifically, pitch is to sound as color is to light. When the light of a distant object is moving towards us, the spectral lines should become compressed, so moving slightly to the blue end of the spectrum (blue-shifted). Conversely, when a luminous object is receding from us, its spectral lines are stretched out towards the red end of the spectrum (red-shifted).

Many astronomers doubted one could ever detect such changes in the spectra of stars. After all, one would have to measure changes as small as a fraction of a millionth of a millimeter! Yet, using his 8-in. Clark-Cooke refractor, Huggins announced in 1868 that he had indeed detected such a tiny shift (about one Angstrom (1×10^{-10} m) in the hydrogen F line of Sirius' spectrum. The Dog Star, it turned out, was receding from us at an astonishing velocity, of the order of some 20 miles per second!

Fig. 9.6 Sir William Huggins (1824–1910)

Finally, Vesto M. Slipher used the spectra of galaxies recorded with the great refractor at Lowell Observatory to elucidate the first extragalactic Doppler shifts (particularly M31, which he estimated to have a blue shift corresponding to a velocity of 300 km/s) used by other astronomers (most notably Edwin Hubble) to unveil the expansion of the universe.

When we stand behind the eyepiece of the classical achromat, we are spiraling headlong towards the grandest of observational mysteries. The extraordinary success of R. G. Aitken, measuring incredibly close double stars with the 36-in. Lick refractor, has already been recounted. Then there's Amalthea, a tiny satellite of Jupiter, measuring just 250 km at its widest extent, which was detected visually by E. E. Barnard on the fateful night of September 9, 1892. Subsequent observations made by Tyler Reed using the 23-in. refractor at Halstead Observatory and G.W. Hough using the 18.5-in. instrument at Dearborne Observatory confirmed Barnard's sensational discovery.

The unveiling of Amalthea was the last visual discovery of its kind to be made in astronomical history and hurtled Barnard to the lofty heights of immortality. Orbiting the giant planet every 11 h or so, this feeble 'spark' of the 14th magnitude shines nearly five million times fainter than its 'primary,' hugging its parent world just a few tens of arc seconds away. If that telescope had so much chromatic aberration (not to mention the attending glare) so as to render it useless, as some authorities have suggested, how could Barnard have detected it so convincingly?

The classical achromat was successfully used in nearly every conceivable climate in which it was used. Its seemingly magical ability to deliver high quality results must, of course, have a sound basis in physics. In this section of the chapter, it shall be demonstrated that these giant pencils pointing towards the sky probably produce the most stable images of any telescope, including modern apochromatic refractors. To see why, read on.

Current wisdom suggests that refractors, i.e., lens-based optics, serve up diffraction limited images most quickly, especially in comparison to Newtonian reflectors or compound telescopes. This author's collaboration with optical theorist Vladimir Sacek, creator of the excellent online optics resource (http://www.telescope-optics.net) has recently identified a number of features that help stabilize the image in small, classical achromats, and which would confer advantages over their faster f-ratio siblings. These are embodied in a widely read online work entitled *Stranger Than Fiction* (see http://www.cloudynights.com/item.php?item_id=2529). The most important findings from that essay are listed below:

- Relative immunity to errors of design
- Generally better e line correction
- Lower sensitivity to focus inaccuracy due to seeing-induced best focus shift
- A reduction of displaced energy around central maxima – the novel Sacek effect
- Higher elevation of the objective above the ground, avoiding ground (and body) turbulence.

As the table below illustrates, making a decent long focus doublet achromat is considerably easier to execute well in comparison to a shorter focus ED refractor. This is particularly true of small refractors of classical design (f/15 relative aperture) (Table 9.1).

The Sacek effect, so named after its discover, Vladimir Sacek, is particularly noteworthy. It is illustrated by the broken red line in the figure below. While performing the diffraction calculation, Sacek discovered that high quality classical achromats have significantly more encircled energy within the first diffraction ring of a stellar image. This results in a reduction in the intensity of the rings seen around the Airy

Table 9.1 Shows errors induced by deviations from design in an f/15 achromat in comparison to a f/6.3 doublet apo of the same aperture (Courtesy of Vladimir Sacek)

Error Induced by Deviations from Design, D = 100 mm			
Design parameter	f/15 achromat induced error in e-line 0.025 wave RMS	f/6.3 doublet APO induced error in e-line 0.050 wave RMS	Tolerance ratio achr/apo
R2 mm (% radius)	5.7 (1.06)	0.17 (0.092)	33.5 (11.5)
Conic	−0.07	−0.0055	12.7
Lens separation	+4.1 mm (from 0.2 to 4.3 mm)	+0.16 mm (from 0.13 to 0.29 mm)	25.6

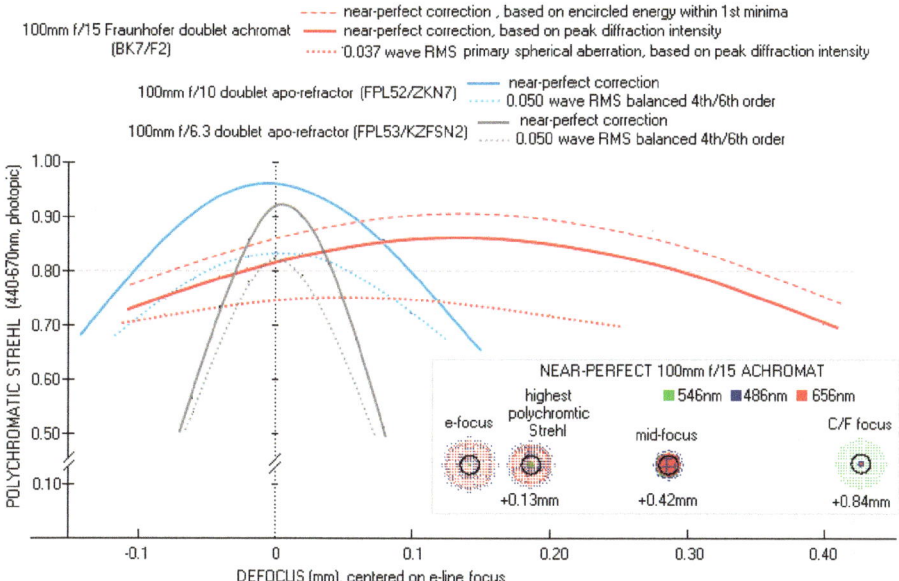

Fig. 9.7 Shows the polychromatic Strehl as a function of linear defocus for a variety of 100 mm refractors of various focal lengths (Diagram courtesy of Vladimir Sacek)

disk, which in turn makes them harder to see. Rings that are less intense are pushed around less by the prevailing seeing conditions, and so the stellar image appears steadier to the eye. In addition, weaker diffraction rings render faint close companions easier to pick off under critical, high magnification tests. What's more, even tiny focusing errors, which are much more likely to occur in faster f-ratio 'scopes than in instruments of slower f-ratio, will throw additional energy into the diffraction rings, rendering faint stellar companions even harder to resolve (Fig. 9.7).

Make no mistake about it:

Stranger Than Fiction applied only to optical systems that have already attained thermal equilibrium with their environments. But what about instruments that are in the process of acclimating, or indeed, reacting to temperatures that are changing? There is considerably more to unveil regarding the thermal properties of refractors.

Glass Facts

Optical glass varies considerably in its ability to expand and contract when experiencing a temperature change. Indeed the coefficient of thermal expansion of these glasses ranges from between 4 and 19×10^{-6}/K. Dr. Juergen Schmoll, an astronomer and instrument scientist based at the Center for Advanced Instrumentation, Netpark, Durham, UK, said that the thermal expansion of low dispersion glasses is significantly higher than either of those used in a classical achromat. Consider the coefficients of thermal expansion (CTE) in tried and trusted crown and flint glasses:

F2: 8.2 * 10^{-6}/K
F5: 8.0 * 10^{-6}/K
N-BK7: 7.1 * 10^{-6}/K
N-BAK4: 6.99 * 10^{-6}/K

Now compare these values to modern low dispersion glasses:

S-FPL51: 13.1×10^{-6}/K
S-FPL53: 14.5×10^{-6}/K
Fluorite: 18.9×10^{-6}/K

The higher the CTE, the more the glass is likely to change shape while acclimating, which in turn affects the definition of the image. For example, a lens that morphs as it cools will be more difficult to focus accurately, as it will introduce aberrations similar to spherical aberration into the optical train. As you can see, the new, synthetic fluorite glasses have CTEs that are ~1.75× to 2× higher than the old glasses, with fluorite itself exhibiting even higher values (~2.5×). This is the reason that oil spacing had been invented for lenses such as the legendary Zeiss APQ series (now sadly discontinued) and those more recently offered by TEC (USA) and CFF (Hungary).

This is a very significant revelation, as plate glass is well known to change shape while cooling. Originally plate glass was employed to make Newtonian mirrors but was gradually replaced by Pyrex, owing to Pyrex's lower CTE (4×10^{-6}/K compared to 9×10^{-6}/K for plate glass). Fluorite and its synthetic derivatives have CTEs roughly double that of plate glass!

We can conclude, with absolute certainty, that modern low dispersion glasses will undergo significant changes in shape as they struggle to acclimate to the outside air, and indeed will continue to change shape as temperatures fall during a typical night's observing. Curiously, the classical achromat, with its continued use of traditional glasses (crown and flint) fares considerably better in this regard. The relative 'hardness' of its constituent glasses ensures that it maintains its figure better, explaining the many reports from amateurs who have noticed that they require less frequent focusing while using them in the field (Fig. 9.8).

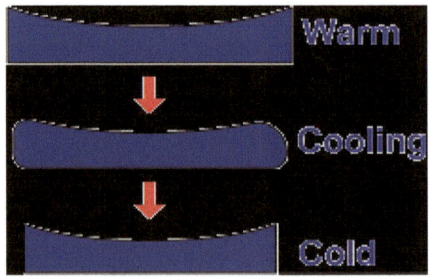

Fig. 9.8 Glass distorts as it acclimates (Image credit: Oldham Optical UK)

Indeed, we have already heard in our exploration of Zeiss refractors the testimony of the Czech particle physicist and avid amateur astronomer, Dr. Alexander Kupco, who posted his findings comparing an older, long focus (f/15) Zeiss AS 80 and a modern Stellarvue SV80S f/6 triplet apochromat. He reported that the long focus doublet Zeiss gave sharper, more stable images than his short tube triplet apochromat right from the beginning of his observing session, and that despite having acclimated, the Zeiss *always* maintained an advantage in this regard.

Lens Thickness and Cooling Rates

The focal length of a simple lens can be determined from the lens maker's formula:

$$\frac{1}{f} = (n-1)\left[\frac{1}{R_1} - \frac{1}{R_2} + \frac{(n-1)d}{nR_1R_2}\right]$$

where

f is the focal length of the lens,
n is the refractive index of the lens material,
R_1 is the radius of curvature of the lens surface closest to the light source,
R_2 is the radius of curvature of the lens surface farthest from the light source, and
d is the thickness of the lens (the distance along the lens axis between the two surfaces).

You can see from the equation that the thickness of the lens d is inversely proportional to f, the focal length. Thus, lenses with long focal length can be made (and generally are made) more thinly than their shorter focal length counterparts. The equation also shows that the focal length scales directly as the radius of curvature of the lens, implying that as R increases so, too, does focal length.

Of course, this is first principle of optics, and it can be modified to accommodate two or three lens elements, but the broad result is the same. After all, an objective is designed so that all the elements *behave as one*, or as closely as possible anyway.

Data supporting the lens maker's formula, particularly the notion that the larger the radius of curvature of the lens the less massive it is, was difficult to come by, but one curious correlation for a series of 6-in. refractor objectives was found for the current run of Istar achromatic doublets:

f/5: 2.65 kg
f/8: 2.6 kg
f/10: 2.5 kg
f/12: 2.2 kg

The 'unwarping' of the lens as it struggles to equilibriate with ambient air temperature manifests itself as a number of aberrations in the image, including spherical aberration and defocus. The author came across this paper authored by J.H. Burge at the University of Arizona. See http://www.optics.arizona.edu/optomech/Fall09/Notes/dfdt.pdf.

Specifically, the rate of change of focus with respect to temperature scales proportionally to f ratio. Specifically, $\Delta f/ \Delta T = \beta$ f, where $\Delta f/ \Delta T$ = rate of change in focus with respect to temperature, f = focal length and β is a constant that only depends on the CTE of the objective glass. Thus, in an idealized system, an f/5 optic will take three times longer (all other things being equal) to serve up diffraction limited images than its f/15 counterpart and thus will suffer from poor apparent seeing for longer. OSLO analysis appears to confirm this generalized idea. The f ratio connection is also alluded to by J. B. Sidgwick in his book, *The Amateur Astronomer's Handbook* (p. 191).

In other online discussions, this author recalls one amateur being astonished at the weight difference between a large (200 mm) triplet apochromat compared to his 9-in. classical Clark objective. Indeed, the mass difference was over 50%! Such an enormous mass differential will have significant results in the field, with the latter achieving thermal equilibrium considerably faster under typical observing conditions.

The connection between focal length, lens curvature and thickness is one of the keys to unlocking the mysteries of classical achromats, and it has been entirely overlooked by modern telescope makers. It is almost certainly responsible for a good part of their magic in addition to that which has already been highlighted. Apochromatic lenses, on the other hand, are usually thicker than achromats. The former are usually triplets, or doublets with a strong radius inside. You can see that when you look into a two-lens ED refractor – the steep curvature between the two lenses is quite striking.

According to Dr. Schmoll, this should affect cooling in ED/fluorite refractors in two ways, once, as the lenses must be thicker to accommodate the steeper radii, so that it takes longer to cool down. On the other hand, the steepness itself means that the difference between the thickest and the thinnest point of the lens is larger, giving rise to a larger dimensional difference during cool-off, and this becomes visible owing to the strong refractive power of the steep lens surfaces.

The Advantages of Depth of Focus

The findings in this author's *Stranger Than Fiction* essay alerted readers to the advantages of depth of focus, and its reciprocal, the defocus aberration, in combating the deleterious effects of seeing-induced focusing errors. The slower (higher f ratio)'scope has a larger depth of focus over its faster (lower f ratio) counterpart and so enjoys a broader range of focus positions over which the Strehl is acceptably high when seeing error subsides. This is clearly illustrated in the figure shown earlier. The faster 'scope enjoys less latitude in this capacity. One can readily see this

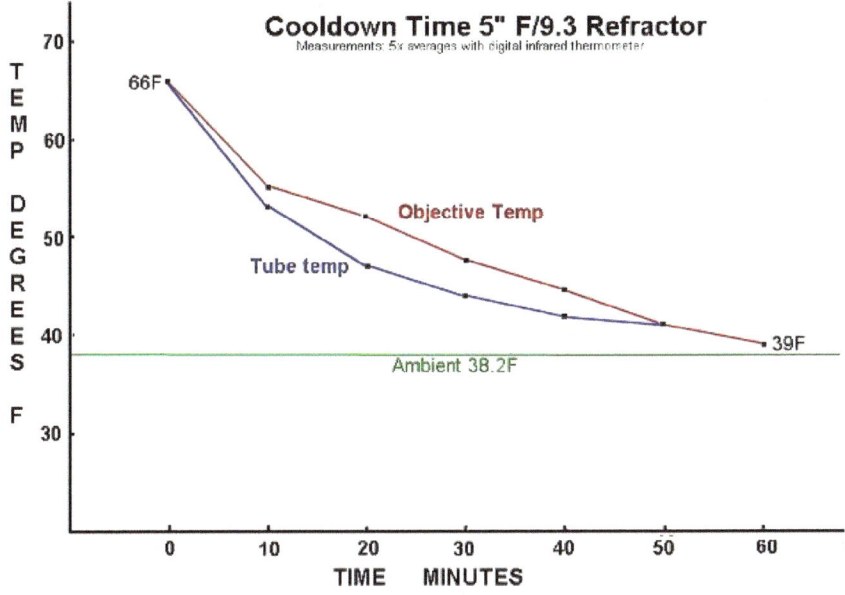

Fig. 9.9 Graph showing cooling curves for a 5-in. f/9 doublet refractor (See http://www. cityastronomy.com/cooldown.htm (used with permission))

effect by hooking up a high f-ratio 'scope and a low f-ratio instrument of the same aperture to a CCD camera. By focusing on the screen, it is easy to see that the high f-ratio 'scope has a greater range of focus positions over which the image remains usable in comparison to its faster f-ratio counterpart.

Cooling-induced defocus, in and of itself, is nothing new. But how does an f/5 system differ from an f/15 instrument as it cools? To see what can happen, consider the depth of focus of the two 'scopes. In the absence of any spherical aberration, the diffraction limited defocus range is given by $4.13\lambda F^2$ and this results in a defocus tolerance of +/− 0.028 mm for the f/5 'scope, whereas the f/15 instrument has nearly an order of magnitude more tolerance at +/− 0.247 mm. Most telescope tubes are made from duralumin, an aluminum alloy with high tensile strength. Suppose you were to set up an f/5 and f/15 refractor at the same time and leave them to cool off. Suppose further that after 15 min or so, you focus both 'scopes as accurately as you can and then leave to grab some coffee. When you returned a few minutes later would you notice a difference? Most certainly!

The CTE for aluminum is 2.3×10^{-5}/K, so the focus shift caused by a change in tube length for, say, a 3 K temperature differential would be 0.104 mm for a 1.5 m long tube, and 0.035 mm for a 0.5 m tube. This tube contraction would place the f/5 'scope outside its allowed defocus latitude, causing the observer to refocus. In contrast, the f/15 image would still be in focus! (Fig. 9.9).

Bearing in mind that it takes at least 50 min for even a modest 5-in. achromatic objective to reach the same temperature as its tube for a temperature change of just 15 K (converted from the above graph), it is reasonable to conclude that larger apertures (with their larger bulk glass mass) will take significantly longer to fully acclimate. What is more, the cooling time is obviously accentuated still further by larger temperature gradients, which can often be experienced during the winter months in cold and temperate climates.

Many decades ago, the great French mirror maker, Jean Texereau, concluded that a temperature difference of less than 1 K within a telescope's tube could degrade the optical wave-front enough to push the instrument outside its diffraction limit. This is not only true of reflective optics but refractive systems also.

Seeing Beyond the Purple Glare

It is obvious from this analysis that smaller telescopes cool off more quickly than their larger counterparts and that refractors have several advantages over their reflective counterparts. But not all refractors are created equal. Air-spaced triplet apochromats usually have their low dispersion element sandwiched between two other elements, which insulates the former and slows its acclimation. Indeed, according to Wolfgang Rohr, a respected optician and tester of optics, a 10-in. air-spaced triplet apochromat would be overkill:

> I have seen one 10-inch f/9 triplet from a commercial maker and it showed no visible color error. That is not to say, however, that the telescope gave a stellar performance. On the contrary, by introducing air gaps back into apochromatic lenses – which inevitably show strong internal curves – we bring back the old problems of the Zeiss B and Taylor triplets, namely their great sensitivity to temperature and to internal alignment. I was rather aghast to see the severe spherical aberration in the 10-inch lens, due to the falling temperature that night. Because of the great thickness and mass of the lens, as well as the fluoro-crown's very high coefficient of thermal expansion and its insulated position in the middle of the lens, this $40,000 extravagant objective never performed as well that night as a decent 10-inch Newtonian would. My impression is that the owner found this true on other nights as well and lamented that the lens could not keep up with the falling temperature.
>
> Other examples of this type of instrument also show the same problem, I am told by my optical acquaintances. So while the smaller lenses of this type in the 160mm range may be fine, it would appear to me that the makers of the larger lenses have overreached the limits of what triplet apos are capable of, at least the air-spaced variety. It is a shame that oiling, the revolutionary technical advance introduced by Wolfgang Busch and Roland Christen almost 30 years ago, has been abandoned. Oiled lenses even of rather large thickness show much more moderate variation of spherical aberration during cool-down in my experience. Perhaps the large air-spaced beasts will work well on tropical islands where the diurnal temperature variation is minimal. But people who live in temperate climates may wish to be careful of large air-spaced ED lenses.

This study has implications for the design of large refractors. For example, building large aperture (>10 in.) apochromats using the less efficient thermal properties of ED glass might not be the best way forward. The lower lens mass and CTE values of crown and flint glass would almost certainly deliver more stable images than their ED counterparts, especially under conditions where temperatures continue to fall throughout an observing run. Color correction could be achieved retro-focally, either by employing a tri-space (examples of which have been built successfully by Roland Christen) or by employing a Chromacorr.

Horses for Climates

In summary, the advantages of high f ratio in an achromatic format are several fold:

- Faster cooling
- Greater image stability due to high entrance pupil, glass properties and depth of focus
- Minimal Seidel aberrations
- Greater image scale for double stars and or other activities requiring precise measurement.
- Minimalist eyepieces that can be used for maximum contrast.

If the thermal data presented above is to be taken into consideration, then it is clearly the classical achromat that has the best thermal properties of all, i.e., it has the lowest rate of change of Strehl of all refracting telescopes and is thus best equipped to deal with changing temperatures in the field. These findings go a long way to explaining why our telescopic ancestors did so well using these simple glass prescriptions of yore. Furthermore, the data also explain why the classical achromat could be used so productively – even at relatively large apertures – across several continents the world over.

Traditionally, the classical achromat has been the instrument of choice to discover and measure the orbital aspects of double stars. This study not only lends credence to that sentiment, it confirms it beyond all reasonable doubt. Furthermore, there is every indication that the classical refractor, using traditional, low expansion glasses found in crown and flint, together with the properties of depth of focus, single it out as *the* telescope best equipped to undertake such measurements. Its rapid and complete acclimation, even under the harshest of conditions, makes it an ideal instrument for the dedicated student with a trained eye. This resonates well with the testimonies of many earnest observers' experience in the field. Indeed, in the time-honored words of Agnes Clerke, who was referring specifically to the classical achromat:

Refractors have always been found better suited than reflectors to the ordinary work of observatories. They are, so to speak, of a more robust as well as a more plastic nature. They suffer less from the vicissitudes of temperature and climate. They retain their efficiency with fewer precautions and under more trying circumstances. Above all, they cooperate more readily with mechanical appliances and lend themselves with far greater facility to purposes of exact measurement.

Fig. 9.10 Long live the classical achromat! (Image credit: Phil Jaworek)

Albert Einstein once said that the most beautiful thing one can experience is the mysterious. And though the properties of the classical achromat are no longer clandestine, they retain their singular beauty as iconic scientific instruments. Truly, they are monuments to human genius, as fundamental to our civilization as are great art and literature. They are veritable magic flutes that have played their sweet notes across the centuries.

Alas, most of us cannot see beyond the purple glare. We are guilty of taking them for granted, and, as a consequence, we have depreciated their utility in the mind's eye. The sad reality is that all too often, like sleeping giants, they sit in great, domed cathedrals that are slowly crumbling away because of lack of interest or funding. Doubtless, many others have been dismantled for parts and will probably never see the light of night again. Needless to say, it is this author's fondest hope that they will continue to play their sweet tunes for amateur astronomers in the decades and centuries yet to unfold (Fig. 9.10).

Chapter 10

Pioneers of the New Glass

The improvement of the achromaticity of the refracting telescope had its origins in the eighteenth and nineteenth centuries, when curious minds sought to devise better lenses for their telescopes. Leonhard Euler, whom we met in connection with the early development of the achromatic refractor, experimented with liquid lenses, employing a zero power meniscus to hold it in place. In Scotland, the naval surgeon Robert Blair investigated the refractive powers of various salt solutions, and between 1827 and 1832, the English mathematician Peter Barlow constructed various liquid-filled lenses in apertures ranging from 6 to 8 in.. His efforts showed that the technology was not only viable but indeed could be competitive with conventional glass-based systems. In the end, though, the development of new glass materials led to the abandonment of liquid lens research.

It was at the end of the nineteenth century, through the extraordinary efforts of Ernst Abbe of Zeiss, Germany, and H. Dennis Taylor of T. Cooke & Sons, England, that refracting telescopes with reduced chromatic aberration in comparison to a standard achromat were produced, using new types of low dispersion glasses. They worked superbly. Indeed, the finest view of Mars this author has ever experienced was with an f/18 Cooke-Taylor photo-visual triplet. Yet, as innovative as they were for their time, they failed to capture the imagination of the astronomical community in general.

Undoubtedly, part of the reason for the failure of the new apochromatic refractors to catch on lay in the extraordinary success of the Newtonian reflector in the latter half of the nineteenth century and first half of the twentieth century. The new glass mirrors, with nicely figured paraboloids and finished in silver, were hard to beat in terms of the bang for the buck they offered in comparison to the more expensive refractor. The serious amateur was often seen to invest in a large 6- or 8-in. Newtonian rather than put his modest disposable income into a 3-in. refractor.

N. English, *Classic Telescopes: A Guide to Collecting, Restoring, and Using Telescopes of Yesteryear*, Patrick Moore's Practical Astronomy Series, DOI 10.1007/978-1-4614-4424-4_10, © Springer Science+Business Media New York 2013

Yet, notwithstanding the obvious superiority of the Newtonian over the small refractor in terms of sheer light-gathering power and resolution, there continued to exist a subset of amateurs and professionals alike who expressed a preference for the more stable and aesthetically pleasing views delivered up by refractors. Moreover, their hassle-free temperament continued to endear them to an army of observers across the world.

As we saw earlier, one of the pioneering companies that continued to invest valuable time and resources into the design of apochromatic lenses was Zeiss Jena. Using a variety of abnormal dispersion glasses, they were able to produce reasonably affordable apochromatic and semi-apochromatic refractors but still retained the high f ratios (typically f/15).

Eastern Promise

The story of Takahashi had its beginnings in 1932, when Kitaro Takahashi founded a sand casting factory on the outskirts of Tokyo. After the desolation of World War II, his factory switched to making aluminum parts for optical instruments.

Realizing the profit margins to be gained from constructing whole telescopes in house, Takahashi decided to try his hand making his own optics, and by 1967 he had succeeded in bringing to market his first refracting telescope, the TS65 refractor (doublet achromat 65 mm f/14). By 1969, the company had produced its first 65 mm triplet semi-apochromat as well as a 100 mm f/10 reflector. This was followed in 1972 by the TS80, the first triplet apochromat ever built. The company had developed ways of artificially growing calcium fluorite (CaF_2) in the laboratory, a synthetic mineral that, as we have previously seen, could provide superb color correction when mated with a suitable element. This TS 80 was used to photograph the total solar eclipse of June 29, 1973, in Africa. It was an 80 mm × 1,200 mm and was offered with a sturdy equatorial mount with a built-in polar telescope.

In 1977 Takahashi introduced its second triplet fluorite apochromat, the TS-90, a 90 mm f/11 instrument, followed in 1979 by two state-of-the-art mounts, the 1990s and the now legendary JP mount. In the same year, the fluorite Series FC 65, 78, 100 and 125 were introduced. This series remained in production until 1994 (Figs. 10.1 and 10.2).

The FC series had uncoated ED elements resulting in a rather strong daylight reflection from their objectives. Given that these elements were apparently hand figured, one at a time, in situ, by optical technicians that had never previously worked such soft glass with these steep curves, the company apparently had a difficult time getting the figure on the ED element right and smoothly polished and didn't want to take any chances altering the figure or roughening the surface with its then-current coating formulations and technology.

After the discontinuation of the FC series, Takahashi introduced its superlative line of FS apochromatic refractors. These were f/8 fluorite doublets that had antireflection coatings applied to all elements and offered in apertures of 78, 102,

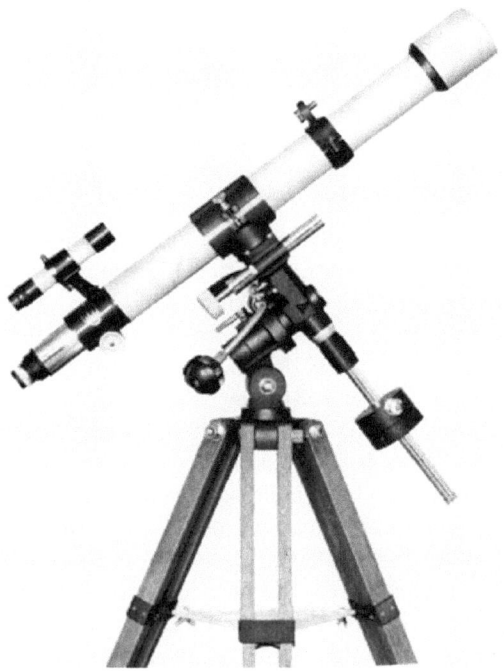

Fig. 10.1 The Takahashi TS90 apochromatic refractor (Image credit: Takahashi)

Fig. 10.2 The strong reflection from the uncoated low dispersion element of an early Meade 127 ED refractor (Image by the author)

128 and 152 mm (3–6 in.). Though now discontinued, they are highly regarded among visual astronomers today.

Meanwhile in the United States, a newly minted optician, Roland Christen, had been, initially in his spare time, busy developing telescopes and accessories for the advanced amateur since 1975. Christen published an article in the October 1981 issue of *Sky & Telescope*, entitled "An Apochromatic Triplet Objective." This was a groundbreaking article and is widely considered to be a cornerstone of the new age of apochromatic refractors that were soon to come. He displayed the prototype, a 5-in. f/12, at the Riverside Telescope Making Conference. The images of Jupiter apparently impressed the attendees beyond any telescope in the field and won the prize for the most innovative optical design. When his first ads appeared in *Sky & Telescope* magazine in December 1981, they caused a great stir among the amateur community. Two objectives were offered: a 6-in. f/11 magnesium fluoride coated oil triplet for $1,800 and an 8-in. f/11 for $3,600, both housed in a well-designed cell.

It is important to note that Christen did not invent the oiling process. Other telescope makers had conceived the idea long before this. The late Horace Dall was using oil-spaced doublets in the 1950s. The oiling reduced the number of air to glass surfaces, thereby increasing the transmission; also, if the oil was of the correct refractive index the objective then exhibited the characteristics of a single block of glass, eliminating any residual figuring errors on R2 and R3. It also removed the insulating air space between the elements, thereby enabling the object glass to cool down more quickly.

Fortune smiled on Roland, as he was able to find a large supply of an abnormal dispersion flint, similar to the Schott KzFS-1 that was originally bought by NASA but never used and which he promptly bought up when it was offered for sale. This glass, when properly mated to other elements, promised even better color correcting properties. Indeed, it is said that no flint glass produced before or since could match its abnormal dispersion properties. Encouraged, Christen produced new ads featuring two new apochromatic 'scopes. A 5-in. f/6 was offered for $950, and a longer, 5-in. f/12 Super Planetary for $975. Unlike his earlier offerings, these were complete optical tube assemblies, utilizing a BK-7/KzFS-1(NASA)/BaF-10 or BaK-1/KzFS-1/BaFN-10 glass prescription.

Next up was the 6-in. f/9 NASA triplet, of which (it is said) only about 24 were made. In the July 1984 issue of *Sky & Telescope*, it was offered at a slightly high $1,695 for that time. For another $1,300, you could have a complete mount and custom tripod. Astro-Physics then expanded their line to a 4-in. f/6 ($795) and 5-in. f/6 ($995). The last model in the line was a replacement for the 6-in. f/9 NASA triplet, the 6-in. f/8 ($1,295), which used KzFSN-4 as the abnormal dispersion flint element. The color and spherochromatism of this model was slightly less well corrected than the older 6-in. f/9 but was $400 cheaper.

Christen continued to hold down a job while making his telescopes in his spare time. He finally crossed the Rubicon, having been encouraged by his wife Marj and his friend and telescope maker Fred Mrozek (of APOMAX fame). Fred had already built and brought to market a limited run of superb 5.2-in. and 8-in. apochromats, and with Fred's father Chester Mrozek (who supplied tooling, a grinding/polishing machine and other important ideas), Roland went into the full-time telescope-making business.

More ads were taken out in the astronomy magazines, and a new, more complete line of apochromatic refractors was introduced. The August 1986 issue of *Sky & Telescope* listed seven apochromatic refractors, the most impressive of which was the 6-in. f/12 SuperPlanetary offered for $1,540. The ad read: "Our new long focus refractors are designed for the most discriminating Lunar/Planetary observer who does not want any compromise in performance… The Lunar limb and the disks of the planets are sharply outlined against black sky, resembling charcoal drawings."

Dave Novoselsky, an amateur and collector of fine apochromatic refractors, said this about his Astro-Physics 6-in. f/12 Superplanetary. "This is the finest planetary 'scope that one could ask for; in side-by-side observing, it has shown more detail on Saturn and Jupiter than a 16-in. Zambuto truss 'scope of focal length 1,800 mm, or the newer Astro-Physics 6-in. f/9 and even a 216 mm f/6 Maksutov Newtonian. I have seen no color with it on any object except Venus, not even on the limb of the Moon. More diffuse objects require a longer focal length eyepiece, but this is not a problem; the Double Cluster in Perseus is absolutely stunning with a 35 mm Panoptic" (Fig. 10.3).

Christen figured (correctly it seems) that there would be greater demand for shorter focal length refractors that could allow wider fields of view for both deep sky viewing and shorter exposures for astrophotography. They would also be easier to mount and so would be more portable. In this capacity, he introduced the new 'Starfire' triplet, a 5.6-in. f/7 Starfire in 1986, followed soon thereafter by 6- and 7-in. f/9 instruments (Fig. 10.4).

What makes Astro-Physics refractors so special? Well, the telescopes were designed and manufactured in a state-of-the-art facility, so the objectives were (and still are) 100% American made. Christen only used the finest "A" grade optical glass, free of striae and other imperfections for maximum transmission and definition. His designs are computer-optimized based on the particular melt charac-teristics of the new glass, allowing his opticians to adjust the tooling accordingly to achieve the desired curves.

All lenses are polished on pitch and hand-corrected. Each lens is tested, polished and re-tested throughout the production process. This continues until the desired performance is achieved. During the final figuring stage, the lens is evaluated with a laser interferometer. Mass-production techniques are not employed, as it is impractical; each lens is treated individually. This process is very time-consuming, but there is virtually no other way to attain the level of resolution, definition and contrast that the most demanding applications require. The combination of the apochromatic lens design, careful, precise optical production techniques, high-transmission multi-coatings and well-baffled tube assemblies result in an instru-ment that will deliver the optical goods when conditions are right.

In August 1987 Astro-Physics relocated to a larger facility that was capable of meeting the demands of a growing number of customer orders. And ever since, waiting times have lengthened. Indeed, some recent buyers have had to wait over a decade to receive their instruments after placing their orders. By the end of the

Fig. 10.3 The Astrophysics 6-in. f/12 Superplanetary refractor (Image credit: Richard Day)

1980s, improvements in computer-aided lens design software and better glass prescriptions led to Astro-Physics introducing their innovative 'Starfire EDF' series of triplet apochromat telescopes. In December 1989 the first Starfire EDF telescope, the Astro-Physics 6-in. EDF, an f/7 triplet apochromat, was produced with a totally redesigned 4-in. diameter field flattener and focuser. Astrophotographers Tony and Daphne Hallas made some beautiful, full-color astrographs that Christen used to promote his products.

The Astro-Physics 4-in. f/8 StarFire – an oil-spaced apochromatic triplet – was manufactured between 1987 and 1991. The earliest models came with an attractive blue tube, but later models were entirely white. The name was changed to "StarFire 102" about 1990. The earliest models had a 2-in. focuser made by a third-party company. Sold without a case, it could be had for around $1,200 (Fig. 10.5).

Fig. 10.4 The oil-spaced triplet objective on the 6-in. f/12 Starfire has weathered well after all these years (Image credit: Richard Day)

Fig. 10.5 The highly regarded Astro-Physics 800 equatorial mount (Image credit: Richard Day)

Fig. 10.6 The Astro-Physics Star12 on a homemade alt-az mounting (Image credit: Jeff Morgan)

In the 1980s and 1990s, Astro-Physics optical designs continue to innovate with each new design, new levels of performance were achieved. The quality of construction of the tube assemblies, sophistication of the mounts and range of accessories have also improved year upon year.

Although most amateurs associate Astro-Physics with triplets, the company also churned out a limited run of doublet ED refractors (Fig. 10.6). Jeff Morgan, a telescope maker and avid observer based in Prescott, Arizona, was kind enough to share his experiences regarding his recent purchase of an older 120 mm doublet Astro-Physics refractor offered between 1990 and 1992:

> When one thinks of Astro-Physics the first thing that generally comes to mind is "triplet apochromat." Yet in the early years A-P made several outstanding doublet telescopes. One of these was the 120 mm f/8.5 Star12. I was fortunate enough to obtain one in December of 2010 and have been using it extensively this observing season. While I am not an optician I will share what I have learned about the history and design of the model, as well give you my impressions of the Star12 observing experience.
>
> The Star12 uses a Steinheil flint leading configuration with KzF2 mated to Ohara FPL-51. The flint-leading design protects the more sensitive (and expensive) ED element, which was left uncoated in the Star12 design. Also, the harder flint element can better resist scratches from cleaning. Cool down is reputed to be faster, and sensitivity to de-centering is reduced. Had Astro-Physics decided to keep the Star12 in production, perhaps the final blow would have been the discontinuance of KzF2 due to environmental problems.

Apparently, none of the replacement materials had the high corrections and internal consistency Roland Christen required.

Although it is not unusual to be giddy and enthusiastic about a new 'scope, the experience with this 20-year-old veteran continues to impress. The Star12 tube weight is 11 pounds, comparable in feel to today's 4-inch apochromats but offering performance comparable to a 5-inch apochromat. After being outfitted with a 2-inch diagonal and eyepiece, finder scope, and Telrad the working weight is still only around 16 pounds, so lifting it to the saddle plate of a tall tripod is quite manageable. Even on evenings where I am feeling a bit fatigued, the Star12 is easy and fast to bring into action. (Don't tell my doctor, but I was observing with it two days after shoulder surgery).

The mechanical features of the 'scope are well designed and executed. The original rack and pinion focuser is still a top performer. The lens cell is adjustable, and mine shows no wear marks; quite possibly the lens is still in factory alignment. My high desert location commonly has a day-night temperature swing of 35 degrees F, but lens cool down has yet to be a problem.

When talking about apochromats, the central issue is always "color." The Star12 specification was a focus variation of 0.07% (1 part in 1430) over the r to h wavelengths (706 nm to 405 nm). I suspect most Star12s are used visually, but correction in the visual C-F range was not published. Astro-Physics claimed the 'scope was capable of focusing the important wavelengths into the Airy disk. To test this I decided my first target would be Vega. After close observation I wasn't really sure I could see any purple fringing without a reference comparison. I was looking so intently for ten minutes I failed to notice Vega's companion! A few weeks later it occurred to me that an interesting experiment would be to observe a number of bright stars, noting the spectral class of the star and the color compared to a reference reflector. The first attempt was 1st magnitude Regulus. Almost immediately I realized it was another fool's errand and abandoned the survey idea. The bottom line is that even though designs have improved over twenty years, Roland Christen saw fit to label this 'scope "apochromat," and I do not disagree.

In terms of figure, the lens leaves little to be desired. Focus has a decisive 'snap' with no ambiguity, capable of working at (far) over 50× per inch of aperture. Delicate airy disks or arcs can be seen even with below average seeing with great symmetry in the diffraction rings on either side of focus. Perhaps my best observation to date is Xi Scorpii. This is a close double that varies from 0.2 to 1.2 arc seconds over 46 years. At the time the separation was about 0.9 arc seconds and increasing. The Star12 cleanly separated the AB components at 254× in Pickering 5 conditions, in agreement with the theoretical limit of 0.94 arc seconds for this aperture. And compared to 4-inch refractors the 120 mm aperture has noticeably better light grasp, which is especially apparent on globular clusters. The longer focal ratio is friendly to the simpler eyepiece designs that are favored for the observation of low contrast features.

Overall I am very enthusiastic about this 'scope, and could only let it go for a larger A-P doublet. If you see a Star12 on the market, I would not hesitate to put an offer on it.

In the late 1990s, Astro-Physics also offered even smaller travel 'scopes for the discerning amateur astronomer on the move. Prominent among them were the Astro-Physics Traveler, a 105 mm F/6 triplet apochromat and, at about half the size, the 90 mm f/5 'Stowaway.' Even by today's standards, these instruments are so well thought of that their price tags on the used market have, until very recently, appreciated (Fig. 10.7).

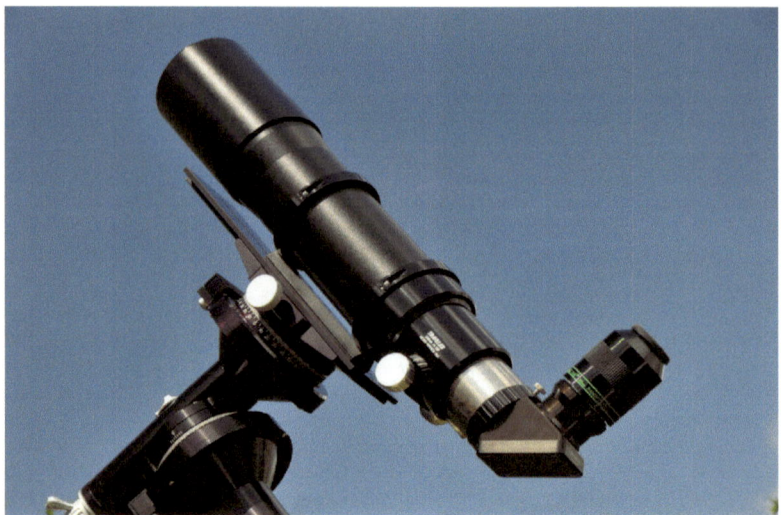

Fig. 10.7 A modern classic – the ultraportable Astro-Physics Traveler EDFS triplet apo
(Image credit: David Stewart)

Round about the same time as Astro-Physics began offering their Star 12ED line of refractors, Meade Instruments Corporation introduced an attractive line of high performance ED doublet refractors (Fig. 10.8).

The all-new Meade 'scopes were drool-worthy all right, stylized as they were astride a nicely designed LXD 600/650 mount. And the user testimonies, in retrospect, were intriguing, too. It turns out that earlier models – produced in 4-, 5-, 6- and 7-in. f/9 formats – had some quality control issues. Several users reported an annoying tendency for the rearward ED element in the objective to become de-centered, necessitating frequent re-adjustment. Other folks complained that it wasn't a true apo, 'just a well corrected achromat' (Fig. 10.9).

The star test on a Meade 127ED conducted by this author was excellent. It showed only very slight under-correction when fully acclimated. Color correction is very impressive. But for the record, first magnitude stars such as Vega and Capella show a very mild purple halo around them in sharp focus.

The Meade 127ED is a first-rate double star instrument. Propus (Eta Gem), Theta Aurigae, Eta Ori and Mu Cygni were easy with this telescope, when conditions allow. It also makes light work of Pi Aquilae, a difficult 6th magnitude pair with a separation of 1.4 arc sec. Where the split is marginal in a 4-in. telescope, the Meade renders much more convincingly. It also serves up fine, high-contrast images of the major planets under good conditions. The best news of all, though, is that this modern classic has been known to go for less than $1,000 on the used market! Grab one if you can!

Fig. 10.8 The Meade 127ED is a terrific performer (Image by the author)

Fig. 10.9 Still performing well after nearly two decades of use, the Meade 127ED objective (Image by the author)

Al Nagler's Wonderscopes

While Astro-Physics and Takahashi were busy improving their triplet designs, Al Nagler of TeleVue Optics traveled an altogether different road. He set out to create the ultimate portable 'scope with enough aperture to keep you going as a visual observer for years, while also delivering the finest flat field astrographs the hobby could yield.

Nagler enjoyed a love of amateur astronomy from an early age, but as a profession, he chose optical engineering. In the 1960s, Nagler became involved in the design of the NASA lunar landing simulators, and it was while working on this project that the young man from the Bronx in New York hit on a way to combine his hobby with his love affair with the sky. Nagler, or 'Uncle Al' as he is affectionately known, sought to deliver to amateur astronomers that seemingly limitless vista created for the astronauts. The first widely acclaimed instrument to leave the TeleVue workshops was the innovative Genesis refractor.

This four-element modified Petzval design was a 4-in. f/5 refractor that delivered improved color correction as well as a beautifully flat field for astrographic applications. But it was a visual observer that arguably made the TeleVue Genesis famous. In the early 1990s, then *Sky & Telescope* contributing editor Steven James O' Meara wrote a book on the Messier objects as seen through the Genesis from his dark sky site on the Big Island of Hawaii. Indeed, he followed up this study by a more ambitious book project that cataloged the visual aspect of the Caldwell objects using the Genesis refractor.

This author has been enormously fortunate to have owned and used an early 1990s Genesis refractor. It's a superbly designed instrument, with a heavy duty, powder coated aluminum tube and a black retractable dew shield (earlier models had a white dew shield) with a beautifully machined threaded lens cap. At the other end you'll be greeted by a chromed rack and pinion focuser. Up front is a 4-in. crown and flint doublet – with a huge air space between the elements. Further back is a two-element, sub-aperture 'corrector' with one of the elements made from fluorite (Fig. 10.10).

Daylight views deliver crisp, high-contrast images. There's a little color at moderate magnifications (75× and above), so it's definitely not an apo by modern standards. Indeed, the chromatic aberration is roughly equivalent to that exhibited by a fine 4-in. f/15 achromatic doublet, so definitely present but not intrusive. But the quality high power views of planets and double stars it served up when pushed to magnifications of 150× or so is impressive. Current wisdom attests that a 4-in. aperture ought to take 200× before the image breaks down, but when this author pushed the 'scope to these higher powers, the planetary images were found to be a little bit soft. Indeed, the views of Jupiter through the Genesis and a traditional 4-in. f/10 Tal 100R achromat were quite comparable, but the longer focal length Russian achromat was the easy winner in terms of sharpness, despite showing more chromatic aberration.

Fig. 10.10 The original TeleVue f/5 fluorite Genesis (Image by the author)

Still, the Genesis could do something the Tal simply couldn't. Insert a 31 mm Nagler 'hand grenade' eyepiece, and you'll get a whopping 5° field at 16× with pinpoint stars right to the very edge of the field. That's a field area four times bigger than that presented by the Tal with the same eyepiece! Inexpensive 2-in. eyepieces fare worse, though, exhibiting noticeable field curvature and astigmatism towards the periphery of the field.

When the Genesis was compared to a 4-in. f/9 ED refractor, a noticeable contrast difference was perceived between the instruments. Scrutinizing the Perseus Double Cluster riding high overhead one cold February evening, the ED doublet just seemed better in this respect, with a darker sky background and more pinpoint stars. The better optical figure and coatings with fewer elements in the optical train of the doublet probably played their part to manifest such an impression. Despite these deficiencies, the Genesis approximates the perfect all-around instrument, built to last several lifetimes and ready at a minute's notice to sail the starry archipelago on a simple alt-azimuth mount.

The Genesis was a great success for Nagler, especially in America, where it has become a well loved modern classic. But Nagler didn't rest on his laurels. He refined the design by introducing improved, low dispersion glasses into, first the front, then the rear elements leading first to the Genesis SDF (f/5.4) in 1993, followed fast on its heels by the TeleVue 101 in 1996. Finally, in August 2001 TeleVue unveiled their flagship 4-incher, the venerable Nagler-Petzval (NP) 101.

These early pioneers of the new glass helped make the refractor popular again. Indeed, their efforts led directly to the 'real' apo revolution that began in 2004, when Synta brought to market an 80 mm f/7.5 ED doublet refractor at a bargain price, allowing many more amateurs to enjoy the aesthetic views these new glass prescriptions promised under good conditions and which in turn was responsible for the enormous increase in refractor sales over the last decade or so.

In the next chapter, we'll be taking a look at another class of telescope altogether, the ultracompact catadioptric instruments that changed the face of amateur astronomy, for better or worse, ever since they first hit the amateur market some three decades ago.

Chapter 11

Classic Cats

Are you after a classic instrument with real, albeit modern pedigree? Are you one of those classic 'scope lovers who desires the finest optics fused with impeccable mechanics delivered in an ultraportable package? Are you after a telescope that you can proudly hand down to your kids? If you answered yes to all three questions, then chances are you have a Questar telescope on your wish list.

Though the business has declined in recent years, Questar Corporation is a privately owned firm that has tasted the sweet smell of success for over half a century. Long considered a world leader in the development and manufacture of premium quality Maksutov Cassegrain telescopes for the amateur astronomer, when you purchase a Questar you are, to some extent, buying into the philosophy that the finest things come in small parcels. For some, Questars are quite simply the finest personal telescopes ever made.

Questar had its inception back in 1950, when Lawrence Braymer established the company to develop and bring to market the highest quality Maksutov Cassegrain optics ever made. The Questar Standard telescope – the one we all recognize and love today – actually began production in 1954 and created quite a sensation when it was first offered for sale in a *Sky & Telescope* advertisement. It will come as quite a surprise to some that Questar does not, and never did, produce its own optics. The hand-aspherized mirrors were commissioned out to Cave Optical, but for most of the company's history the optics were produced by Cumberland Optical.

Questar soon found markets for their optical wares in many and diverse niches, including the military, the police, security, and in the aerospace industry, as well as the consumer market.

The most familiar instruments sold by Questar include a 3.5 and 7-in. (the Q3.5 and Q7, respectively) aperture Maksutov Cassegrain astronomical/terrestrial telescopes for the consumer market. But while it was produced in very limited numbers, Questar also once offered observatory class Q12 optical-tube assemblies,

N. English, *Classic Telescopes: A Guide to Collecting, Restoring, and Using Telescopes of Yesteryear*, Patrick Moore's Practical Astronomy Series,
DOI 10.1007/978-1-4614-4424-4_11, © Springer Science+Business Media New York 2013

which were sold along with a state-of-the-art equatorial mounts based on a Byers drive system. Even NASA was lured into purchasing a few Q12s for visually tracking their rocket launches.

Braymer began working on the design of his Questar 3.5 as early as 1946. His goals were clear-headed and exacting. His Questar would offer up top-notch images in a package that was at once easily transportable and easy to use. Braymer settled on a Maksutov Cassegrain design for the tube assembly, named in honor of its inventor, the Russian-born Dmitri Maksutov (1896–1964).

Maksutovs come in a variety of flavors. The Gregory type is the simplest, using all spherical surfaces. The secondary mirror has exactly the same curvature as the corrector. This design is used by Meade and Celestron, because it can be mass produced. Then there's the Rumak, which incorporates a separate secondary mirror, which has the advantage of a separate curvature to fine tune any residual aberrations left on an all-spherical Gregory-type Maksutov.

Braymer chose neither, however. Instead, his Questar would embed the secondary on the same glass as the corrector, but it is ground to an independent, aspheric curvature. The secondary has an aspheric curvature, figured by hand. This is a very expensive process, but it does arrive at a very convenient result – the secondary can never go out of alignment. The aspheric element is claimed to provide superior optical performance, but this has been disputed by some amateurs. Braymer used a modified Cassegrain design that added an aluminized spot to the Maksutov corrector plate, creating a compact folded light path, hence the f/14 relative aperture of the instrument.

By placing the spot on the inner (R2) surface of the corrector, Braymer knew that he would be copying the patent design owned by John Gregory, which was already licensed to Perkin-Elmer. So instead, Braymer placed the spot on the outer (R1) surface of the corrector lens in the earliest models. But by the mid-1960s, the patent issue having been settled, Questar Corp. began to adopt the Gregory design with the aluminized spot on the inside of the corrector for extra durability.

Intriguingly, the original blueprint for the Questar design was envisioned as a 5.1-in. (130 mm) telescope, but Braymer later revised those plans, as the instrument would prove prohibitively expensive for even well-to-do folk.

Every aspect of the design of the Questar, which encompasses some 200 separate parts, was exceptionally well thought out. For instance, Braymer designed a built-in control box that enabled the user to switch between the main telescope and a coaxial finder 'scope by simply looking through the main eyepiece. This is achieved by the elegant simplicity of a flip mirror. Conveniently, this also allowed the user to mount a camera or other auxiliary devices. Other notable features included an exceptionally smooth focus knob and a built-in 2× Barlow lens. The cast-aluminum double-fork arm mount was designed with a built-in Right Ascension (RA) clock drive and could be used in equatorial mode simply by adding the collapsible legs (also included).

The Questar 3.5-in. entered commercial production in 1954, and almost immediately this 'mobile observatory' was hailed as the "Rolls-Royce" of telescopes. Ads for the model have run in all the glossy magazines over the years, including

Fig. 11.1 The ad showcasing the 50th anniversary limited edition Questar (Image credit:
Questar Corp)

National Geographic, *Scientific American* and *Sky & Telescope*. The Questar of the
1950s and early 1960s offered little capacity to employ third-party accessories. For
example, the original Questar could only incorporate 0.96″ eyepieces, but soon a
range of accessories, including the now standard 1.25-in. oculars and other accessories, would be made available (Fig. 11.1).

Over the years the company has designed several variations on a magnificent theme:

- A 3.5-in. Field Model Questar, which was the first type offered for sale, in May 1956. This is the most basic and least expensive model to purchase.
- A 3.5-in. Questar Standard, with an integrated star chart engraved on white on a blue aluminum sleeve (which also doubles up as a dew cap) around the barrel, which contains a Moon map.
- A 3.5-in. 'duplex' model, which included an elegant and simple way of detaching the telescope from the fork mount, to enable the optical tube assembly to be used as a f/12 telephoto lens.
- A 3.5-in. distance microscope that came with two barrels: the standard one, and a separate longer barrel, which could be screwed onto the duplex fork mount in place of the standard one, and used as a distance microscope. This variant came with the standard accessories, and a separate case containing the distance microscope barrel assembly.
- The very popular 3.5-in. Questar Birder, which consisted of a modified Questar Field Model possessing a fixed 10× finder with a rapid-focus knob used for observing birds and other wildlife.

In 1967, Braymer introduced a 7-in. (180 mm) model to the market. In essence, it is just a scaled up version of the original 3.5-in. Questar. But if you thought the original model was expensive, the Q7, as it's affectionately known, never made the same impact with the amateur community owing to its extremely steep price tag. Indeed, less than a 1,000 were built by the end of 2011. The majority were used by the U.S. army and other special agencies for long distance reconnaissance. A few were also used by NASA to follow spacecraft during the first minutes after launch (Fig. 11.2).

In 2002, Questar Corp announced that the Q7 had undergone a makeover. Introducing their lightweight titanium 7-in., the company replaced all the internal parts previously fashioned from stainless steel with lower density titanium. That alteration removed some 4 lb off the weight of the instrument in comparison. But perhaps more significantly the move over to titanium improved the thermal properties of the instrument, enabling it to cool off more rapidly in the outside air. Alas, it is unknown whether this made any difference to the salability of the instrument in the long run.

Prior to 1975 Questar made instruments as large as 7 in., but by this time, a stream of new and ergonomic products were beginning to flood the market; an inundation of Newtonian and Schmidt Cassegrains reflectors in 8–10-in. large apertures ostensibly devouring the company's profit margins. Bizarrely, Questar embarked on what, at least in retrospect, must now be considered to be a dangerous strategy. Instead of working towards widening their product accessibility, they embarked on the design of an even larger and more expensive optical system, settling in the end for a 12-in. Maksutov, marketed to applications in industry and government. Notwithstanding its breathtaking price tag, it seemed the natural

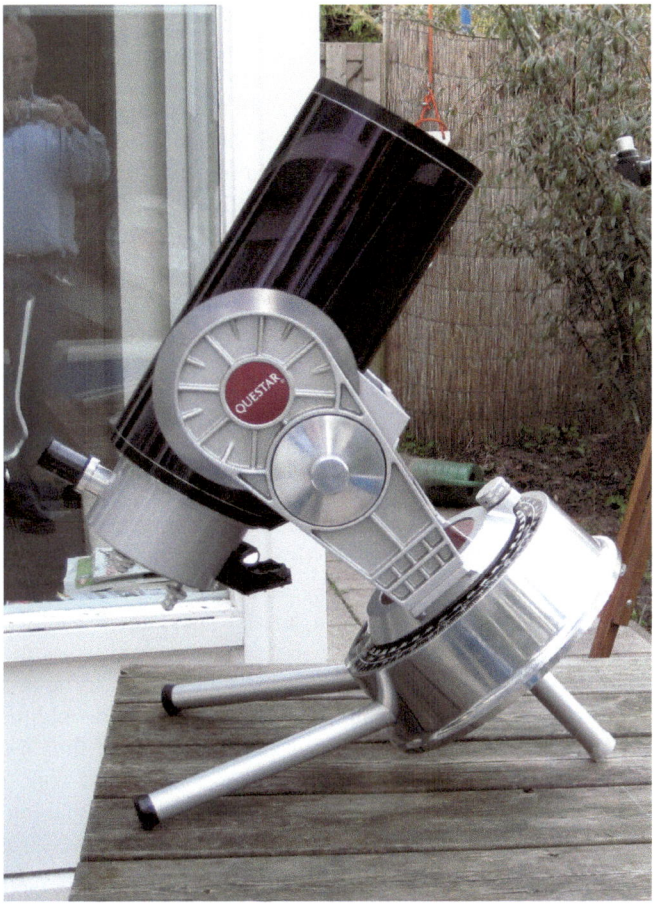

Fig. 11.2 The beautiful Questar 7 (Image credit: Erik Bakker)

choice, too, for the wealthy and discerning amateur astronomer. As it turned out, the production of such a large, complex optical system and the problems that attended its mounting, proved to be extremely labor intensive and time-consuming. After all, the Q12 was to incorporate a precise asphere on the concave side of the BK-7 (borosilicate crown) glass corrector lens, which required extraordinary skill to undertake. Indeed it is said that a battle-hardened technician might expend as much as 6 months of hard graft in figuring, mating the optical components and testing the system. Needless to say, Questar Corp. sold few of these telescopes. Incidentally, a Q18 was also rumored to be in the cards, but the project was apparently (wisely?) shelved owing to lack of sufficient commercial interest.

The Questar 3.5 has gained a solid reputation as an heirloom telescope, even though its diminutive aperture is easily outperformed by larger 'scopes at a fraction of the cost. It is, in essence, a complete mobile observatory that has provided enthusiasts with a lifetime of observing pleasure. NASA astronauts aboard the Gemini and Apollo spacecraft took one into space. Questars have enjoyed a loyal following from many well-known astronomers and non- astronomers alike. Rocket pioneer Wernher Von Braun acquired one in 1959. The late U. S. talk show host Johnny Carson was a well known amateur astronomer and purchased an early model Q 3.5. The founding host of NBC's *Today Show*, the late Dave Garroway (1913–1982), and the science fiction writer Sir Arthur C. Clarke (1917–2008) were also well-known owners. And in more recent times, famed comet hunter David H. Levy also acquired one.

The optical figure of the Questar telescopes is reported to be about 1/10 of a wave p-t-v and comes equipped with two Brandon eyepieces – highly prized in their own right – in 24 mm and 16 mm focal lengths. In more recent times, the substrate used to make the primary mirror of the Questar has been upgraded to incorporate ultra low-expansion glass as an option. The instrument's Zerodur ceramic mirror does away with the annoying need to refocus the telescope as it cools to ambient temperature, especially in situations involving large temperature swings. If the difference in temperature between indoors and outdoors is 30° or more Fahrenheit a Pyrex mirror contracts as it cools and will thus require minor refocusing. Although the 3.5-in. mirror of the Questar should cool down much more rapidly than a larger mirror, it is sealed inside the tube, making cool down slower or more protracted. That's where a Zerodur mirror helps because it under-goes virtually no expansion or contraction as temperatures change. The mirror in this 'scope even eliminates the need for minor refocusing (Fig. 11.3).

Bob Abraham, a professor of astronomy at the University of Toronto and keen amateur observer, provided some considerable insight into why the Questar tele-scope appeals to people:

"Although I've collected telescopes off and on for decades, it's only relatively recently that I felt my bank account would let me get away with picking up a used Questar. While they are not exactly cheap on AstroMart, good ones can be found for under 2 K. While one can certainly get a lot more performance/dollar for 2 K than a used Questar, once one gets a look at the craftsmanship, quality of materials and rather elegant features of the 'scope (like the built-in Barlow) I felt (and feel) 2 K is a reasonable deal. However, to feel this way you have to be the sort of person that greatly prefers owning something with beautifully machined aircraft-quality alumi-num components that cost 10× as much as something made of plastic that does essentially the same thing. Anyway, Questars seem to last forever and don't seem to depreciate very quickly, which also made me feel a little better about buying one.

In terms of performance, all the Questars seem to have excellent optics. I now own two of them (upgraded to a Duplex model with a Zerodur mirror, which is

Fig. 11.3 A classic Questar 3.5 (*right*) awaits nightfall (Image credit: Bob Abraham)

handy in Canadian winters). The optical quality in both Questars was found to be comparable to that of the TEC Maks (my all-time favorite amateur 'scopes). One can have a lot of fun with a little Questar. I'm rather busy with my day job (professional astronomy!) and find that I don't have much time to indulge in my hobby (amateur astronomy!). Between work and clouds I typically only have an hour or two to sneak out once or twice a month to do a little observing. The Questar is so easy to set up it is well suited to this style of usage (provided you remember to keep it stored close to ambient!). The Questar is a wonderful little double star 'scope and a nice quick-look lunar 'scope. If Jupiter is up or Mars is at opposition there's no way the Questar should be a first choice for planets, since planetary views are exactly what you'd expect from a really good 90 mm catadioptric, which in performance is somewhere in between a 60 mm and 80 mm apo. But for casual quick looks at planets it's a nice 'scope. The built-in Barlow is terrific and has rather unusual finder. If you live in a city and find stuff with setting circles -- the Questar's circles are large and easy to use – you will not miss a go-to at all.

Other things to like about the Questar: it's a very pretty telescope that doesn't take up a lot of room. You can keep it on your desk and admire it almost as a work of art, and occasionally pop outside with it for a look at birds and insects from your backyard. This highlights its portability. It is really fabulous to see it emerge from such a small and elegant case, and is trivially easy to move around. It would be a great eclipse 'scope.

The 'scope, while being lots of fun, isn't perfect. The forks are not very tall, and you might have trouble getting to southern objects in equatorial mode because the tube would bang into the base. And, rather obviously, the field of view is limited. If you live in a city deep sky observing (aside from double stars) is pretty hopeless anyway, so a 1 degree field of view is all you want. But the finder's aperture is probably too small for star-hopping in a city, and unless you are comfortable with setting circles you might be frustrated finding fainter things in a light-polluted environment. In some ways a casual urban observer living north of 30 N that is comfortable finding stuff with circles is an ideal Questar owner.

Another thing to note: our department owns a few Questars for student use, and while these ones are 40+ years old they're in comparatively good shape, which speaks volumes about the 'scope's longevity. However, one of the department's older Questars is slightly out of collimation and is not fixable by the owner. So they're not totally impervious to abuse and some aspects of the design are not user-serviceable, which is a weakness. On the other hand, Questar does a great job of servicing these telescopes, and there are not too many 40+ year old 'scopes that you can send back to the manufacturer for repair if needed.

Let us conclude with this: at the end of the day, Questars are charming and fun small 'scopes. Their capabilities are limited for what they cost, but they're beautiful, very nicely made and well-suited to quite a few types of observations. They last forever. They are specialized and have lots of character, but lack modern amenities."

Erik Bakker, an amateur astronomer from the Netherlands, has had the pleasure of owning and using a Q7 and kindly provided his opinions on its form and function. "The Questar 7 comes in two large black cases, he said, "one contains the optical tube assembly with eyepieces, solar filter, eyepiece filters, adapters, camera coupling and power cord. The other case holds the fork mount and table top legs. The tube weighs in at around 22 lbs, the fork mount with tabletop legs around 35 lbs. The scope can be carry a few yards when fully assembled. The easier thing to do is to assemble the fork mount first, then put that on a sturdy (picnic-) table. Then you mount the OTA via a massive 0.75 in. bolt to the fork mount, with alignment made easy by two small pins in the fork mount that fit into corresponding holes in the optical tube. All in all, you are ready to observe in 3–5 min! The Q7 is the bigger brother to the 3.5 Questar Duplex. It is a massive scope compared to the tiny and light Q3.5.

The Questar 7 Classic has the famous control box at the eyepiece side of the OTA. It contains the finder, star diagonal, a 1.6× Barlow, 1.25 in. eyepiece port on top of it and an axial port.

Everything in the control box is operated by the flick of two levers. One for magnification (the Barlow) and one for the finder. Both are very convenient and work exceptionally smoothly. The Questar optics are made for Questar by Cumberland and are sublime. I had mine professionally tested and the wavefront of the whole system including the diagonal and Barlow was better than 1/10th wave P-V at the eyepiece with very smooth optics, resulting in a perfect star test. The Questar 7 comes with the Questar Brandon eyepieces, which are a perfect match in terms of contrast, comfort and resolution. Favorites for the Q7 are the 24 and 12 mm Brandon (Figs. 11.4 and 11.5).

Fig. 11.4 Questar 7 compared to 4-in. f/8 fluorite (Image credit: Erik Bakker)

Fig. 11.5 The incomparable 24 mm Questar Brandon eyepiece (Image credit: Erik Bakker)

"I had mine for four years and really enjoyed it. It's a beautiful 'scope, especially on its classic fork mount. And a piece of telescope-history as well. For a 7-inch f/16 instrument it is very compact and lightweight. Construction is superb, really a professional instrument, much more so than a little Q 3.5. Weight and effort to set it up are comparable to a modern short focus 5-inch apo on a German equatorial mount. Very doable and ready to observe in say 3–5 minutes from opening the boxes to observing behind the eyepiece.

In my moderate sea-climate, cooling of the 'scope and optics was not a big issue, but for the best images it does need 1.5 hour or so on most nights to stabilize from room temperature (20 degrees C) to outside temperature (10 degrees C). Longer with bigger temperature gradient of course.

On the older instruments, the focusing is superb, with a very smooth and precise feel. Image shift is practically zero, unlike in the small Q, where it can be plainly visible. The newer Q7 Titaniums have a somewhat coarser feel, but still very nice. The finder is very nice for the Moon and brighter planets. For dimmer objects, it is just a bit too dim, which makes deep sky observing time-consuming due to the long time required to find fainter stuff. Here the TeleVue Starbeam could be a nice addition, as in the current Q7 Astro.

The fork mount is nice and works well for visual observing, for long exposure photography, a modern German equatorial mount is a better idea. Optics are generally really good in the Q7. In vintage scopes, just check the edge of the primary for coating deterioration. And check the corrector coatings for transparency; milky appearance means trouble.

The figure and polish on my Q7 were stunning, easily better than 1/10 wave P-V for the whole system, including the built-in diagonal! If all you did for the rest of your life was looking at the intra- and extra-focal images, you could be a happy camper. I have never seen better optics.

Now after all these good things, here is the trouble with the Questar 7: it has a large central obstruction and a fast f/2 primary. That hurts image quality of low contrast details and in aesthetics of double stars. On the planets, especially Jupiter suffers from that. It takes seeing with Pickering 9 and 10 for the Q7 to show what it can do. Up until Pickering 8, a superb 4-inch apochromat like my FS102 f/8 fluorite doublet will show more detail on Jupiter. Mind you, a little less good 4-inch apochromat, like my Celestron FL102, could never quite reach the Q7, ever.

The bottom line for the planets? The Q7 is a superb and beautiful 'scope. On Mars and Saturn, the same difference is there, though not always as visible as in the very low contrast details on Jupiter. On double stars, with that much light in the first diffraction ring, a Q7 does not show the prettiest image of doubles; here the refractor excels. It does separate to the theoretical limit of it's aperture when seeing allows, showing a hard Airy disk with a bright first diffraction ring on most stars.

On the Moon, a Q7 is very nice and less handicapped by its central obstruction. Very nice and color free views, with minute detail visible. Children and astronomy-laymen love the Questar 7 for its looks, convenience of observing and what it shows of the moon and planets. Especially Saturn is a great match for the Questar 7.

All being said, the Questar 7 is a truly wonderful, timeless instrument, as long as you take good note of what its strengths and weaknesses are. If you buy a well cared for used Q7, you invest your money in an instrument that will last well into the next century. Depreciation and maintenance will be very low, and service from Questar is easily obtainable for U. S. residents at reasonable cost."

The Working Man's Questar: The Meade ETX 90RA

In 1996, Meade Corp. introduced a revolutionary new telescope that sent shock waves around the global amateur community. Called the ETX, it was marketed as an obvious poor man's Questar. Like its more famous cousin, it sported top notch optical performance but at a price that wouldn't break the bank. The ETX RA was fork mounted in a frame made from ABS plastic instead of stainless steel. Three user-supplied AA batteries inserted on the underside of the mount base allowed smooth tracking of celestial objects in right ascension (Fig. 11.6).

The ETX is a 90 mm Maksutov-Cassegrain design with a focal length of 1,250– 50 mm shorter than the Questar 3.5 – and compressed into a tube less than a foot long. And like the Questar, it is fork mounted with a right ascension drive built into the base. Three small legs are included that provide an equatorial mount on a table top, which cater for latitudes as high as 66°, which pretty much includes everyone.

Although you can attach the base of the 'scope to a heavy duty photographic tripod for terrestrial viewing, but at 9 lb, it pays not to skimp on sturdiness. More recently, Meade has also offered the ETX in spotting 'scope format, that is, without

Fig. 11.6 The little ETX RA ready for action perched on your garden table

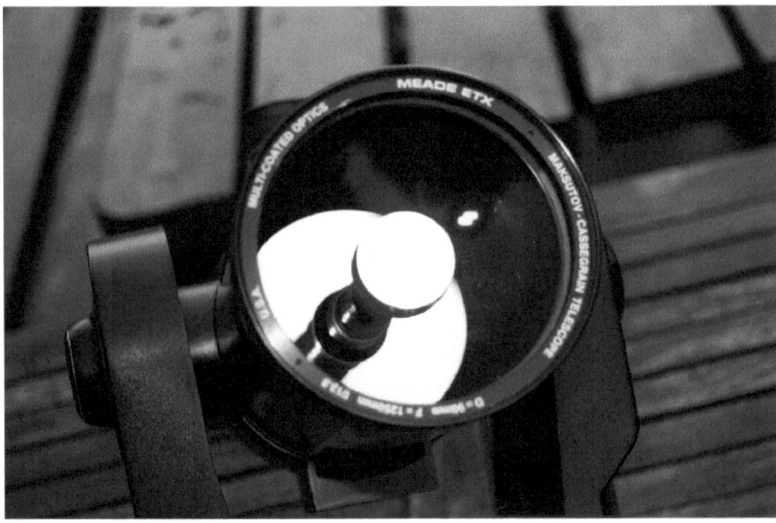

Fig. 11.7 The Maksutov Cassegrain design of the ETX offers up impressively sharp images when properly collimated (Image by the author)

the base and fork mount that can work well on a sturdy camera tripod. The tube contains a flip mirror that allows either straight through viewing or viewing through the top-mounted eyepiece holder. An erect image prism diagonal can be used for terrestrial viewing although it is advisable to switch to regular viewing mode (using the built-in mirror) for high power applications (Fig. 11.7).

Installing the batteries involves removing three screws and detaching the base plate using a screw driver. You must set the hemisphere in which you are observing to either north or south for the drive to work properly. When switched on the motor makes a very slight humming noise. The motors can take up to a few minutes to take up the slack in the worm gears, and some observers deem it advisable to let it run for at least 5 min before doing any serious tracked observing (Fig. 11.8).

The instrument came equipped with some standard accessories, including a 8×21 viewfinder and a 26 mm super Plossl eyepiece to get you observing more or less immediately. The finder is, sadly, next to useless. Indeed it should come with its very own health warning, as it is excruciatingly difficult to view through, especially when the telescope is pointed near the zenith. Meade cannily responded by offering a right-angled finder as an optional upgrade, but most users think it represents only a marginal improvement over the original.

Focus is achieved by rotating a knob that moves the primary mirror either towards or way from the front corrector plate. Like almost all such systems, there is a little bit of slop, meaning that the image is slightly shifted when one approaches focus from opposite directions. The declination slow-motion control is only usable when the drive is unlocked, so you have to rely on the drive alone to keep an object

Fig. 11.8 The advertisement featuring the original ETX RA (Image credit: Meade Instruments Corp)

in view if you wish to use the motor. The declination, on the other hand, has a very effective slow motion control that works after the declination is locked (Fig. 11.9).

The beauty of a Maksutov Cassegrain design at this aperture is that all surfaces, including the corrector plate, are spherical, and so the Meade ETX has been able to achieve consistently high quality optics. All 90 mm ETXs that this author has looked through have wonderfully crisp optics, when properly acclimated. It really is a fantastic little lunar and planetary 'scope. Because of its small central obstruction and very effective baffling it delivers high contrast images that render very similar performance to a similar-sized refractor.

How have these 'scopes aged over the years? To find out, this author purchased an ETX RA for a reasonable price on the used market. It was soon discovered that the front baffle – placed just inside the aluminized center spot – had slipped and thus was de-centered from the aluminized spot. The result was an instrument that was slightly out of collimation, producing slightly oval-shaped stars just inside

Fig. 11.9 The rear end of the ETX houses the focuser, a photo-port for attaching cameras, and a built-in flip mirror (Image by the author)

and outside focus. Since the 'scope was out of warranty, an attempt was made to re-center the baffle by carefully peeling it off the inside of the corrector plate (which is glued on). After applying some fresh adhesive, the baffle was re-centered on the aluminized spot. Star testing revealed a better functioning instrument.

After opening up the underside of the instrument and applying fresh batteries, it was discovered, rather disappointingly, that the instrument failed to track objects. Clearly, this particular unit was badly in need of a servicing. Notwithstanding these issues, the views were nice, both at low and high power, once the instrument had fully acclimated. Certainly, they were never intended to last forever, in a way the Questar perhaps is. But in introducing a low cost, high quality telescope to the masses, Meade Instruments can be duly congratulated.

The Orange C8

The Questar 3.5 and the ETX 90, marvelous though they are, are very small instruments. Of course, they'll show you everything that a 3.5-incher will show. But to go deeper into space and to see finer details on the Moon and planets, you need larger aperture. Until the 1970s, the only viable option for many amateur astronomers was to make or purchase a large Newtonian telescope or invest in a long, heavily mounted classical refractor. But optical advances throughout the twentieth century allowed for the possibility of creating a compact, large aperture telescope. In 1932 Bernhard Schmidt of the Hamburg Observatory began experiments into the design of a new kind of photographic telescope with a wider field of view and a very low focal ratio.

Schmidt hit on the idea of using a thin lens, more correctly called a *corrector plate*, placed in front of an easy to make spherical mirror. This corrector plate served to eliminate spherical aberration, an optical defect inherent to spherical mirrors. The image plane, set about midway between the corrector and the spherical mirror, would of course, be highly curved, but by placing photographic film on a curved mounting plate, pinpoint star images could be achieved across the entire field of view.

After demonstrating the success of the Schmidt camera on several large telescopes, optical designers soon realized that one could adapt the design for visual use. This could be achieved by placing a spherical mirror at the center of the front corrector plate, which would direct the focused light through a central hole in the spherical primary mirror. By carefully designing the secondary spherical mirror to have a given radius of curvature, a substantially flatter field of view could be attained; enter the Schmidt Cassegrain Telescope (SCT).

Some optical firms produced large, custom made SCTs to order. These were very expensive to build and very difficult to manufacture. But by marrying new computer optimized designs with readily available raw materials, it was, unsurprisingly, an electronics firm that first brought the commercial SCT to fruition. Beginning in 1954, Tom Johnson (1923–2012) and Alan Hale of Valor Electronics made various prototypes, before settling on the classic 20 cm (8-in.) f/10 SCT, which saw first light back in 1966. That telescope embodied what amateur astronomers came to know as the orange C8. It marked the beginning of a new revolution in amateur astronomy that continues unabated even to this day. The SCT offers large aperture in a compact format, enabling its users to engage in visual as well as photographic projects. It was truly the first 'jack of all trades' commercial telescope.

The original, or "classic," Celestron C8 SCT had standard aluminum coatings on the mirror and no coatings on the corrector – unless they were marked "Special Coatings." Along with the optical tube, the only standard accessories included a single medium focal length eyepiece. The diagonal and finder 'scope were still optional accessories. By 1984, the orange tube Celestron C8 had been replaced with a shiny black model dubbed the Celestron C8 SPC. Significantly, the company began offering the telescope on a beautifully designed for mount and heavy-duty tripod called the Super Polaris.

These early models were held in place by large rings around the optical tube that in turn connected to the equatorial head. As time went on, the design was upgraded to a bar called a *dovetail* – a type of sliding connection. Both of these configurations allowed for moving the tube into the proper position to re-balance when accessories – heavy and light – were added. With a strong background in electronics, Johnson and Hale added a computer system to the mount replete with a 450-object database of pre-selected items that could be found automatically once the date, time, latitude and longitude were known. Measured by today's standards of computer-guided tracking mounts, it was rather primitive, but at the time of its inception it was truly a marvel of electronic engineering.

In 1984, the Celestron C8 "Super C8" came along with further design innovations on the mount. The drive base shape was changed to square to incorporate a Byers 359 tooth worm gear drive with a single synchronous AC motor. Other design modifications included an 8×50 helical finder 'scope that allowed for either straight-through or right angle viewing – as well as a finder 'scope with a diagonal and an illuminated reticule polar finder 'scope. An additional eyepiece was added to the telescope package, and the wedge system was beginning to advance. This is when Celestron's "Starbight" multi-coatings began to be offered as an option!

In 1987 the Celestron C8 advanced again to CompuStar, an automatic GOTO for over 8,000 objects. The f/10 optical tube with "Starbright" coatings was supplied with an 8×50 polar axis finder, a 2-in. star diagonal and a 50 mm (2-in.) Plossl, heavy-duty wedge and a tripod and carrying case. Starting in 1993 the Compustar was shipped with a 1¼-inch star diagonal and eyepiece with the 2-in. accessories as options. By 1989 the design evolved, and upgraded versions were then called the Celestron C8 Classic 8-in. (Fig. 11.10).

Things got even better when late 1989 saw the release of the Celestron C8 Ultima. The new processor allowed four digitally controlled drive rates, including solar, lunar, sidereal and king. It now worked in both hemispheres just by flipping a switch and introduced periodic error correction (PEC) – the ability to track almost any object and to "train" the drive to automatically counteract the errors that are inherent in any gear system. It improved once again with the Celestron C8+, where the mount became even more simplified and user friendly. This progressed into the Celestron C8 Ultima 2000 line, culminating (at least from a classicist's perspective) with the Celestron C8 CeleStar (Fig. 11.11).

The original C8s were manufactured more or less continuously from 1970 to about 1982–1983. To this day, the Celestron C8 has barely changed its optical standards. If it ain't broke don't fix it, say the SCT fans! The C8 has an aperture of 203.2 mm (8 in.) working at a focal length of 2,032 mm (80 in.) for a well balanced f/10 focal ratio. Capable of gathering 834× more light than the human eye and reaching a limiting magnitude of 14, these spotless optics still provide a minimum of 0.57 arc sec in resolution, putting them up there in a class with a much larger aperture 'scope. Although each model line will be defined by its own set of accompanying letters that denotes the type of mount that comes with the package (for example, the CGE-800 model is a Celestron C8 optical tube on a Celestron CGE mount – or the Celestron C8 CPC-800 with its user friendly features as pictured here), the telescope design remains true to its original orange tube design introduced by Tom Johnson.

Celestron's smaller SCT, the venerable C5, has also enjoyed a loyal following, since it was first introduced in the early 1970s. The bright orange C5 sported a 5-in. aperture and focal length of 1,270 mm (f/10) on a scaled-down version of the fork mount attending the larger C8. The 35-lb assembly mounts securely onto its dedicated tripod, and its inbuilt RA drive tracks celestial objects smoothly once placed in equatorial mode.

One enthusiastic owner of the C5 wrote, "I found this 'scope used at a good price. The motor base is the same as the old C8s with smaller twin fork tines and

Fig. 11.10 The Classic C8 (Image credit: John Leader)

an ac motor that tracks well. Mine came with an adjustable heavy duty field tripod, and with the smaller 5-inch tube it is much more steady than the C8s I've looked through. Stars at high power are clean Airy disks with more light scattered in to the diffraction rings than my TeleVue Ranger. This 'scope gives surprisingly good views of the planets and easily outperforms the Ranger in all areas except for low power wide-field viewing. This particular 'scope is in very good condition cosmetically and has a no nonsense feel to it that makes you want for little else" (Fig. 11.12).

The C5, like its bigger brother, can accept a variety of accessories for visual and photographic applications. Insert a 0.63 focal reducer and you have a fairly capable rich-field instrument to ply the vast expanses of the summer Milky Way. On the

Fig. 11.11 A considerable amount of engineering went into the design of the fork mount on the C8 (Image credit: John Leader)

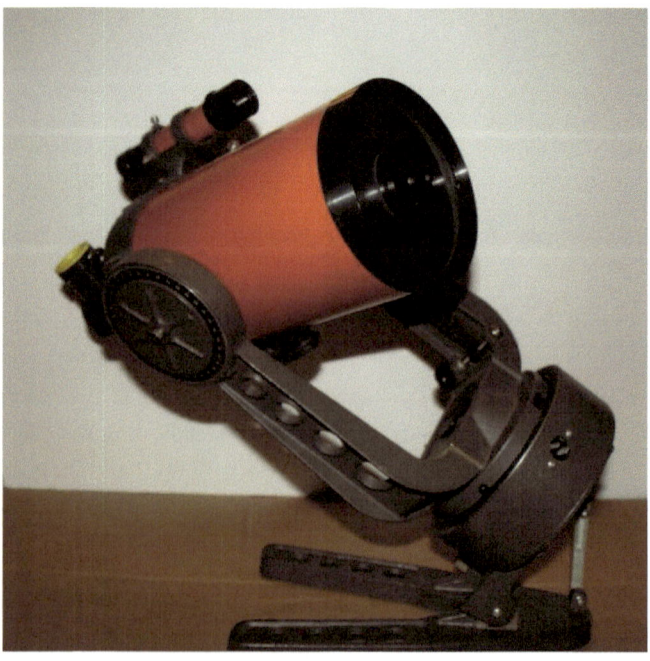

Fig. 11.12 The Classic C5 (Image by the author)

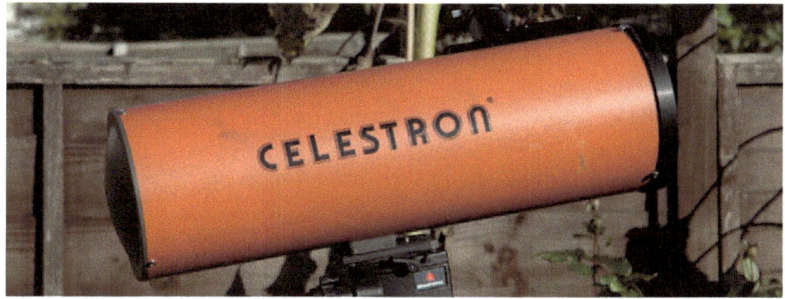

Fig. 11.13 The Comet Catcher basks in the summer sunshine (Image credit: Richard Day)

other hand, if your forte is astrophotography, the exceptional portability of the C5 will enable you to image at either f/6.3 or f/3.3 (by acquiring a 0.33 reducer/field flattener). In that capacity, it is even more versatile than its larger sibling.

Finally, no lover of classic telescopes could possibly fail to mention the Celestron Comet Catcher. This cute little Schmidt Newtonian, with a 5.6-in. (140 mm) aperture and focal length of 500 mm (f/3.6), tipped the scales at only 7 lb and was only 19 in. long, making it an excellent, ultraportable rich-field telescope and astrograph when it first hit the market in the early 1980s (Fig. 11.13).

Owners of the Comet Catcher generally reported decent images up to about 100×, with diminishing returns when pushed to powers much beyond this. Still, the little instrument produces lovely, well corrected wide-field views of a raft of deep sky objects and is ideally suited for quick scans of the heavens, hence its name 'Comet Catcher.' This author has seen these go for as little as a few hundred dollars in the online classifieds.

Meade Enters the Race

They say imitation is the sincerest form of flattery. That's certainly one way to describe the entry of Meade Instruments Corp. into the business of producing commercial SCTs for the amateur market. Company founder John C. Diebel turned his background in electronic engineering into producing a viable competitor of Celestron.

Celestron had the entire SCT market to themselves throughout the 1970s. Founded in 1972, Meade initially sold a small selection of classical achromatic refractors and reflectors, together with a wide range of accessories. It was a humble affair, run for the most part from Diebel's kitchen table. But the man had higher aspirations. By 1977, he could see that the market demand was growing rapidly for compound, portable telescopes. Up to this time, only Celestron had succeeded in mass producing good quality SCT's, even though two other companies tried and

Fig. 11.14 The Meade Model 2080 SCT (Image by the author)

failed in their efforts. Regardless, after a 3-year research and development effort that required all of the engineering and financial resources that Diebel had at his disposal, the company's blue tube Model 2080 8-in. Schmidt Cassegrain telescope was announced in September 1980. Since then, Meade and Celestron have been worthy competitors in designing ever more clever catadioptrics with advanced optics and electronics to boot (Fig. 11.14).

Perhaps in an attempt to imitate its commercial rival, Meade also produced a limited run of small SCTs in a 4-in. aperture; enter the model 2045. With a focal length of 1,000 mm (f/10), the 'scope received mixed blessings from amateurs over the years. Although many have learned to love it for its ultra-compact size and decent optics, others have had less flattering things to say about it, such as: "It has an all metal construction, no plastic junk, but the mount is a bit wiggly. The clock drive uses a stepper motor, but with no speed control. The mount and drive work in very cold weather, but its mirror focuser freezes. Though small, it takes a surprisingly long time to cool down. Collimation is a pain, but that's a characteristic of any SCT."

What is the verdict concerning these telescopes in the field? How do they stack up against other, more established telescopic genres, such as the classic Newtonian or the age-old refractor? Though opinions differ, most experienced observers (your truly included) would opine that the commercially made SCTS are good all around 'scopes but not exceptional. The ferocious competition that ensued between Meade and Celestron in the mid 1980s, perhaps looking to cash in on the return of Halley's Comet to our skies, resulted in many of the SCTs manufactured having shoddy optics, as quality control standards plummeted for maximum profitability. Indeed, it was these mediocre 'scopes that put a lot of amateurs off SCTs for good.

The truth is that SCTs, like all other telescope genres, can be divided into three categories – the good, the bad and the excellent. This author has enjoyed some really nice views of the planets with a few 8- and 10-in. SCTs. They are quite capable telescopes. When they underperform (which has happened too frequently for some) the culprit is either poorly collimated optics or bad thermal management, or both! Thankfully there are enough user reports available online to squarely pin-point SCT models that have deservedly acquired a bad reputation. These include almost any SCT manufactured during the mid to late 1980s, as well as any early model Bausch & Lomb Criterion SCTs from the 1970s. According to catadioptric guru Rod Mollise, "Some are better than others.... But I've got to be honest. I have never, ever run into one that was optically better than 'just passable.' I guess that means that one should avoid them like the plague."

Chapter 12

Resurrecting the Master's Glass

Sir Patrick Moore (born March 1923) needs no introduction, either to the astronomy community or the world at large. An institution in his own right, this genteel, bachelor astronomer for over half a century has graced us with his encyclopedic knowledge of the wonders of the universe and how best to engage with them. But like almost all famous names, Sir Patrick's beginnings were altogether more humble.

His interest in astronomy was piqued as a young boy, aged six. And by 1934, when he was just 11, Moore had saved enough pocket money to purchase his first refractor, a beautiful 3-in. f/12 Broadhurst Clarkson & Co. achromatic refractor. He was guided in his choice by Dr. W. H. Stevenson, a leading member of the BAA, who helped him select the instrument. Costing the princely sum of £7 10 shillings, it rested on a pillar and claw tabletop mount. But as the young astronomer soon discovered, the set up – which he wryly referred to later as the 'Blancmange' – was woefully unstable for serious observation. Thankfully, for an extra 30 bob, he was able to place the instrument on a sturdy wooden tripod.

Moore's choice of telescope was a wise one, in retrospect. Even today, you can't go far wrong with a 3-in. refractor with these specifications. And although today's market is awash with short tube versions with fancy glass, one can still purchase longer focal length instruments not broadly dissimilar to the one Sir Patrick began his telescopic adventures with all those years ago (Fig. 12.1).

N. English, *Classic Telescopes: A Guide to Collecting, Restoring, and Using Telescopes of Yesteryear*, Patrick Moore's Practical Astronomy Series, DOI 10.1007/978-1-4614-4424-4_12, © Springer Science+Business Media New York 2013

Fig. 12.1 In need of a makeover: Sir Patrick's 3-in. refractor prior to refurbishing (Image credit: Steve Collingwood)

Indeed, even with the greater choice available to the contemporary amateur, a 3-in. refractor is still a good choice for a starter 'scope. It can, for example, reveal the main atmospheric features of the giant planet Jupiter, the majestic ring system of Saturn, and the polar caps and deserts of the Red Planet. Many of the brighter deep sky objects come alive in a 3-in. refractor, and an impressive suite of double stars can be readily resolved. Sir Patrick reminded us also how versatile a 3-in. refractor can be in recording the changing aspects of the solar disc, using his telescope to project an image of our star's disc onto a piece of white paper.

The 3-in. objective is an uncoated contact doublet – antireflection coatings were not to be applied to lenses until after the Second World War – made from traditional crown and flint glass. With a focal length of 36 in., it was well corrected for all aberrations, putting up images broadly similar to even the finest contemporary 76 mm refractors. The various components that made up the tube were of very high quality by today's standards.

Sir Patrick, however, had by this time developed a fascination with lunar study, and his 3-in. Broadhurst Clarkson & Co. was put to immediate use in divining the Moon's many wonders. The 3-in. refractor telescope quickly proved its worth in the young astronomer's hand. Indeed, his first paper, "Small Craterlets on Mare Crisium," published in the BAA journal, was based on observations made with it.

Moore was only 14 years old! Indeed, the same telescope was used extensively by Sir Patrick throughout his long career. But after 73 years of field use, understandably enough, it began to look its age. Moore was anxious to have it restored.

Time for an M.O.T

The natural choice, of course, would be to commission the work to the original company. But times, sadly, had moved on. As we have seen, the 1970s witnessed a flood of less expensive Japanese imports into the UK, which conspired with the increasing cost of manufacturing telescopes by hand in such a way as to make a major restructuring of the company a necessity (Fig. 12.2).

In 1973, the company was purchased by Dudley Fuller, who brought with him his successful 'Fullerscopes' line of telescopes and mountings. As well as stocking Telescope House with new product lines sourced from around the world, Fuller had the presence of mind to maintain the traditional methods of telescope production and restoration. Master Craftsman Ernie Elliot, who had started with the firm in 1942, continued to see new commissions, renovating and restoring brass telescopes

Fig. 12.2 The fully disassembled Broadhurst Clarkson & Co. reveals the quality of its component parts (Image credit: Steve Collingwood)

Fig. 12.3 Gleaming with pride – the exhaustively cleaned and polished optical tube (Image credit: Steve Collingwood)

and microscopes for auction houses, antique dealers and collectors. Elliot continued to make instruments using the traditional techniques and tooling (a lot of it more than a 100 years old). And while Elliot sadly passed away a few years ago, the traditional ways of telescope making and restoration he practiced have been kept alive as an important part of the company's heritage and identity. Thus, the restorative work on Sir Patrick's 3-in. refractor was entrusted to Steve Collingwood, Service Manager at Telescope House, who oversaw the project from start to finish (Figs. 12.3, 12.4, and 12.5).

On close inspection, the telescope was in quite bad shape due to the years of wear and tear. The lacquer on the body of the telescope had oxidized and pitted, and there were a few dents, too. Once on the workbench, it became clear that the telescope would have to be completely dismantled and repaired from the ground up. Each individual component (down to every last screw) was carefully removed, repaired as necessary and cleaned by hand – machine polishing would have potentially damaged the now fragile brass. The internal rack and pinion focuser was repaired and the inside of the tube re-painted matt black. The dented lens cell was re-formed as much as possible – again, being careful not to damage the thin brass, and the doublet lens was thoroughly cleaned and re-seated.

As well as refurbishing the telescope, Steve also did a superlative job cleaning up the elegant tripod the instrument was mounted on. He explained:

The real aim of the project was not to actually 'restore' the telescope but to service it and give it a new lease on life. With that in mind, the brass work was carried out with a view to keeping the age of the instrument evident, while ensuring a hard wearing finish to prevent further damage – the dents and scratches make up part of the instruments character.

Fig. 12.4 The tripod before

Fig. 12.5 And after restoration (Image credits: Steve Collingwood)

It is also fair to say that as the instrument was originally craftsman made using traditional techniques and tools, it was important to also show it that had once again been back in the hands of its maker and the same care and skill given over to it.

As you might expect, cleaning up a precision optical instrument such as this requires some specialist skills and great attention to detail. Steve was asked to describe his approach.

"Most of the skill involves knowing how far you can go with a particular part, when to stop and when to leave well alone. I would call it 'reading' the brass, really. For instance, you can't machine polish old brass bodies because it thins the brass too much, leaving it prone to distortion and cracking with the heat generated by the buffer.

It's worth noting that modern methods – if there are any, considering no one makes them in the traditional way anyway – are machine based. Without exception, a telescope maker today will buy the brass tube from a metal yard and turn the ends square on a hobby lathe. They will buy a lens in a cell ready to fit (occasionally they may make a counter cell), the focuser will be brought in from a large supplier and then the thing will literally be assembled by hand. This is both good and bad. It produces a nice telescope, I'm sure, but where is the 'handcrafted' instrument? In the old days, the tubes were made in house from cut sheets hand rolled on a mandrel before brazing (which was done with enough skill that the seam would be hard to spot), the smaller draw tubes were roughly formed, wired, soldered and drawn before finishing, the fittings were cast outside and finished in house using patterns. Lenses were ground and figured by the opticians upstairs, and the brass work was hand polished before applying lacquer by brush.

So what kind of tooling was used in fashioning these old 'scopes?

Generally, they were old iron mandrels that were used as molds for rolling the body tubing, a couple of odd looking small brass drifts and a homemade screwdriver. I still have the old hand drill that was used for all the screw holes on brass bodies. It was so labor intensive and skill-dependent that it simply couldn't survive commercially nowadays. I think that's why it wound down over the years."

From its inception, the project took several months to complete. But it was worth the effort. "Sir Patrick was visibly moved when he saw his 3-in. refractor restored to full health," Steve said. "It even brought a tear to his eye and indeed was a surprisingly moving occasion for all present" (Fig. 12.6).

Steve was also responsible for the restoration a 5-in. f/12 Cooke, which was acquired second hand by Moore. Housed on a heavy duty equatorial mount in a wooden, roll-off roof observatory, the lens of this old refractor was made by the great telescope maker Thomas Cooke, whose legacy was discussed earlier in the book. The tube as well as the lens cell housing the objective are not those made originally by Cooke however, and actually date to the 1960s. Likewise, Steve informed me that the mount upon which the instrument moves was built by Charles Frank. Moore purchased the Cooke objective from Peter Sartori, a wealthy businessman for just a few pounds. The roof of the observatory slides open by turning a handle, linked by a chain to a simple gear mechanism (Figs. 12.7, 12.8, 12.9, and 12.10).

"The lens itself, while largely pristine, had a few air bubbles clearly in view. That's normal for the period," Steve explained. "I had to cut off an aluminum ring

Fig. 12.6 The finished 'scope reunited with its proud owner. *From left to right*: Dudley Fuller, Sir Patrick Moore and Steve Collingwood (Image credit: Steve Collingwood)

Fig. 12.7 A disheveled 5-in. Cooke refractor as it appeared before restoration (Image credit: Steve Collingwood)

Fig. 12.8 The 5-in. Cooke being restored…

to get at the object glass," he continued, "the cell itself could be collimated but the original 19th century brass screws were badly corroded. After cleaning the objective and cell, I had to re-center the lens and bench test it to ensure that the elements were properly aligned with respect to each other."

Fig. 12.9 …and after it was restored (Image credits: Steve Collingwood)

Moore says of his telescope, "It is very good indeed – you won't get a better one!" Indeed refracting telescopes were always his preferred choice in telescopes, serving up slightly crisper images than reflector telescopes.

This superb telescope is still frequently used by astronomers, especially for lunar and planetary observations. Steve said:

"I've spent quite a few nights using it. It's one of the sharpest refractors I've used – and as I spend a lot of my day testing and adjusting optics, that's no

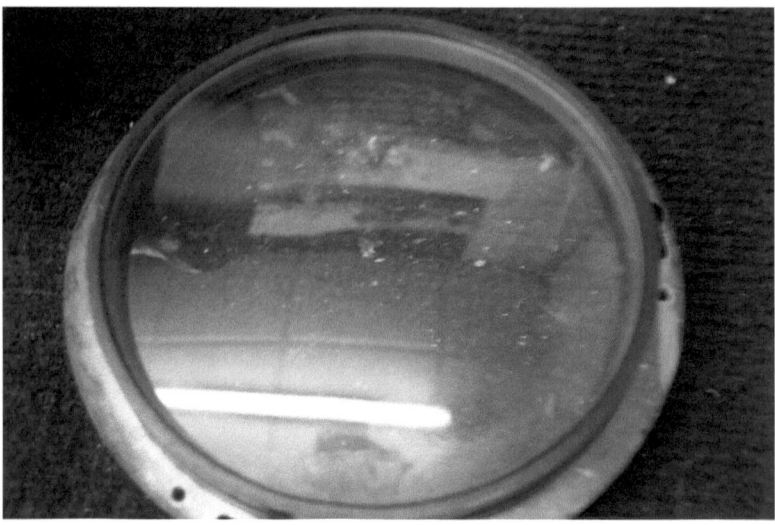

Fig. 12.10 The uncoated, late Victorian objective was in need of extensive cleaning but was otherwise pristine (Image credit: Steve Collingwood)

throwaway statement. It's one of those telescopes that's just. The fact that it belongs to someone as famous as Patrick Moore doesn't really come into it. The physics of this instrument are just spot on. Sir Patrick still swears blind it's a triplet, even though I've shown him that it's a doublet.

I have fond memories of one crystal clear evening at Selsey. A few of us had gathered round the restored Cooke inside the observatory. Eagerly, we took our turn eyeballing Saturn as it crossed the meridian. All you could hear were exclamations of joy. The small wooden observatory was filled with the sounds of 'oohs!' and 'aahs!' Saturn was rock solid in the eyepiece and the details on the globe just 'popped.' Then, suddenly, the door of the old observatory creaked open and the silhouette of a tall, hooded chap appeared in the twilight. 'What have we here gentleman?' a voice cried out. It was a distinctive voice, one which I had heard before but at that moment was caught unawares as to its identity. The man walked over to us, pulled his hood down and said, 'The views are magnificent through this telescope aren't they?' It was none other than Dr. Brian May, the famous guitarist from the British rock legend, Queen."

Chapter 13

The Antique Telescope in the Twenty-First Century

We are nearing the end of this broad survey of the classic telescope market. As we have seen, amateur astronomers derive great joy from restoring and using them under the stars. But old telescopes, like any other avenue of technological history, are also sought out by collectors, who can cough up breathtaking sums of money for an instrument with the right mixture of condition and provenance.

Antique telescopes, as you might expect, are like any other investment – they have their own set of risks and possibilities. Right off the bat, after having canvassed the opinion of many collectors and sellers of antique astronomical equipment, if your intention is to specialize in this area, the consensus advice this author received is simply not to. Most dealers in antiques sell a wide range of items across several genres and so can rely on customers with a variety of interests.

Then there's the problem with any antique items; they need to be stored and protected, as well as insured against theft. If you live in a damp or humid climate, a special room with de-humidifiers is probably a necessity, especially if you have a lot of wood mass. If you have to contend with damp summers and very dry winters, the wood may actually end up cracking on you. Leather has a similar tendency to crumble or rot if not properly cared for. The lacquered brass that is commonly seen on antique telescopes also tends to become discolored over time. One can, however, acquire special waxes to help prevent this, but now you're getting into antique preservation – which can become expensive, not to mention time-consuming.

If you are even slightly unsure of how to proceed or not the handy type, it would be best to have the work done by a professional restorer who can do the job for you. We met one professional resource in the superlative work of Steve Collingwood at Telescope House, London, UK, who rejuvenated two fine classical refractors.

N. English, *Classic Telescopes: A Guide to Collecting, Restoring, and Using Telescopes of Yesteryear*, Patrick Moore's Practical Astronomy Series, DOI 10.1007/978-1-4614-4424-4_13, © Springer Science+Business Media New York 2013

The antique market is as fickle as any other. From 2002 through 2006, antique telescope prices appreciated steadily, owing to the growing popularity of a variety of classical or antique telescopes in the popular literature. With the present economic downturn of 2008 onwards, antique telescope prices have decreased sharply. Indeed this author has seen dealers, in some desperation, offering their wares for less than 50% of the price offered only a few years before.

That being said, aspiring to own a true antique telescope is not something to shy away from. As we have seen, restoring them can be fun, and using them, inspirational! Just be realistic about your purchase, and if you wish to sell it be prepared for ill-guaranteed returns. The finest 4-in. achromatic refractor from the days of yore will not conjure up any more magic than a good contemporary model. Physics is hoodwinked neither by nostalgia or necromancy!

From a collector's perspective, replacing a missing lens element or mirror will more often than not do little to increase your telescope's intrinsic value. Since a discerning collector requires as much of the original piece as possible, warts and all, this will not enhance the value at all. However, if your intended purpose is to use an antique telescope then by all means have the optics fully restored or replaced. As before, entrust the work to a reputable company that handles fragile equipment and antiques or a private individual skilled at doing such work.

Don't be tempted to coat old lenses, either. Most antique objectives are doublets, and so by adding an ant-reflection layer it will only marginally increase light transmission. More importantly, adding coatings will almost certainly depreciate its value to collectors who, almost invariably, prefer authenticity over any perceived improvement.

What's in a Name?

The value of an antique is often dictated by the name of the maker on the tube. You cannot easily claim a telescope is an Alvan Clark if the name is not engraved on the tube. That said, as we have seen, the optics for many lesser names in the industry could also have been made by more famous manufacturers. And indeed, it could work the other way, too. For instance, the Dollond and Cooke dynasties occasionally put their name on microscopes that were made by another maker but sold by the former. One can sometimes ascertain this by studying the engraving style on the objective, but an experienced instrument antiquarian can acquire a feel for the proper engraving style for an era or a particular location.

Imitation is the sincerest form of flattery, or so they say. And human nature being the way it is, folk were quick to cash in on the success of established firms such as Dollond, nefariously marketing look-alike 'scopes under names such as 'Dolland.' Optically, they were probably much the same. Unless formerly owned by some famous individual, the latter instruments, irrespective of their condition, will never have the same pedigree as a *bona fide* Dollond.

The price an antique telescope is able to command depends on a number of things, some subjective and others objective, including the instrument's provenance, condition, age, and whether it's truly unique, unusual or not. For instance, spyglasses that have two or three draw tubes are fairly common. Spyglasses that use a larger number – between four and seven – are much more rare and will therefore command heftier prices. The material used in the 'scope construction makes a difference, too. For example, French-derived unsigned brass telescopes with their common leather covering are fairly widely available and thus not too valuable unless kept in pristine condition along with, for example, an original case and table tripod. On the other hand a signaling telescope covered with more choice material such as ebony wood or whale bone will fetch a much higher price at auction.

Like in any other arena of the antiques market, the higher the stakes the more likely the potential chance of fraud. Before making an expensive purchase, do your homework carefully. Examine the item thoroughly, and if possible, bring along someone more knowledgeable than yourself. Of course, if you plan to observe with the telescope, arrange for the instrument to be used under the night sky before committing to the sale.

Sometimes scams occur with fairly modest instruments. For example, this author was recently alerted to a likely case of lens swapping in a Zeiss Telementor. The unsuspecting buyer paid a premium price for the legendary performance of this small, 63 mm-long focus refractor and after setting it up in his backyard discovered that there was something seriously wrong with the optics. The image had so much spherical aberration that stars and planets could not be focused properly. After testing a few more times, the unlucky owner knew that the lens was a lemon. Contacting the previous owner, the poor chap was told in no uncertain terms that he had not tampered with the lens in any way and that, as a collector, he had merely acquired it in the belief that it was the genuine article.

The new owner removed the objective cell from the tube and was surprised to learn that the usual Zeiss engraving was nowhere to be seen. Suspecting a scam, the chap had the objective tested and, after a sensitive double-pass auto-collimation test, discovered that the objective had a figure significantly worse than the ¼ wave industry standard. As we have seen, this is very unlikely to have been passed by Zeiss opticians, as it was clearly sub-standard. The consensus opinion was that the Zeiss objective had been removed and then replaced by one of poor quality.

The moral of the story here is that fraud can and does happen. When purchasing any expensive item of equipment it is always worth trying it out before committing to it. A small Zeiss telescope, like the one showcased above, ought to perform flawlessly out of the box. It can make all the difference between joy and disappointment.

That said, there is already an astute awareness of the quality of antique and modern classic telescopes, as well as a burgeoning desire among a growing number of people in the amateur community to get involved in acquiring, restoring and using vintage telescopes from the days of yore. And that can only be a good thing.

Outreach and All That

In May 2008, the Antique Telescope Society convened a meeting at Cincinnati Observatory Center, Ohio, to discuss the alarming rate at which astronomical institutions and the instruments they contain are falling into disuse because of funding issues. A prominent member of that society, Trudy E. Bell, summarized the state of affairs thus:

> *Between 1840 and 1940, more than 250 private and institutional observatories housing telescopes and other astronomical instruments of various sizes were founded across the United States, plus a similar number in other nations combined. Many were still functioning well into the 1960s and 1970s. In the latter third of the twentieth century, however, a great number – including some prominent institutions equipped with large, exquisite instruments – fell into disuse or were wholly demolished. Today, scores lie in disrepair and are in danger of being razed. Historians have become so alarmed about the increasing pace of destruction of historically significant astronomical sites and artifacts that several high level initiatives have been launched worldwide to preserve and protect them.*

Source: http://www.capjournal.org/issues/09/09_30.pdf

Sadly, Bell's indictment is as true today as it was the day it was written. But that does not mean that we must sit back and accept the status quo. Getting involved in projects that help restore or preserve old telescopes need not necessarily involve a financial output on the part of the interested person. There are many fine old telescopes across the world that are falling into disuse because too few people have taken it upon themselves to invest the time to learn and evangelize about their former glory days. Many private astronomical clubs and societies both in the UK and in America frequently engage in outreach events to attract new recruits to the hobby. All too often, these well meaning enthusiasts bring along the latest high tech gear, to give passersby a look at some accessible celestial showpieces. But why not bring along a classic 'scope to such events? It need not be a 12-in. behemoth. A small, elegant refractor upon a wooden tripod might be just as appealing to the unbiased eye, or a newly restored Cassegrain reflector on a tabletop mount might be just the ticket for the right person. One could also use that opportunity to tell them something of the telescope's history and how it helped forge the history of astronomy (however small).

If your local astronomical society has an historic telescope lying dormant, pay it a visit and take the time to learn a little about its history so that you can share its story with new members that may know little or nothing about it. There are also a number of societies dedicated to the history of the telescope and, where possible, their restoration and upkeep. Most prominent among them is the Antique Telescope Society (the website of which is cited at the back of the book), an international organization devoted to historic telescopes, large and small. Many local or national science museums contain a wealth of information on astronomical topics and are well worth visiting if you get a chance. The bibliography and websites given at the end of the text will also provide more background for those who wish to explore this fascinating avenue of our wonderful hobby. Good luck with your adventures!

Epilog: Sailing to Byzantium

After the boys were snuggled up safe in their beds one cold December evening, my wife and I took our turns removing the clutter that had accumulated on top of the solid oak case. With a broad smile upon our faces, we lifted the heavy lid to unveil the spyglass inside, a magnificent looking instrument, finished in black and adorned with brass trimmings. I could almost hear the beating heart of that magic flute, decorated as it is with a perfectly executed lens of 4-in. aperture and 5-ft focus.

Like Queen Cleopatra of old upon her litter, together we carried the great little instrument along the landing and down the stairway to an open backdoor. For a lingering moment, it felt as if we were traversing a Stargate, where one world ends and another begins. In passing under that doorway, we knew we were leaving behind, if only for a while, the cares of the workaday world, and entering a netherworld of endless possibility.

With loving care, I secured the 4-in. f/15 in its cradle and removed the cap to let the lens taste the frigid evening air of mid-winter. A cold front had just left our shores, and the waning gibbous Moon, almost full, was rising majestically above the treetops marking our eastern vantage. Mighty Jove was big and bright near the meridian, shining like a distant lamp set inside a window. Vega, sinking fast in the western sky, flickered calmly, while the bright stars of Perseus and Cassiopeia near the zenith were all but steadfast.

My wife prepared some green tea while I dutifully selected an eyepiece. To see the King of the Solar Worlds at his most splendid, I knew that I'd be giving my high power oculars a night off, a brief winter siesta, if you like. I needed to charge the magic flute with a 12.5 mm eyepiece delivering 120 diameters, an enlargement big enough to see all the available detail the instrument was likely to show.

Sighting along the slender black tube, I aimed for the giant planet and zeroed in with the 8×50 finder. Excitedly, I placed my eyeball to the eyepiece and slowly

N. English, *Classic Telescopes: A Guide to Collecting, Restoring, and Using*
Telescopes of Yesteryear, Patrick Moore's Practical Astronomy Series,
DOI 10.1007/978-1-4614-4424-4, © Springer Science+Business Media New York 2013

brought the bright fuzz to an exquisitely sharp focus. And what a glorious sight to behold! Jupiter and its retinue of four large Moons, three of which were arranged in a delicate triangle to the west of the planet, and the other, a tiny, solitary orb, away off to the east.

The globe of that failed Sun was beautifully rendered, tack sharp and predictably steady, stained with numerous tan colored clouds of ammonia and phosphine ice, and sculpted into delightful quasi-linear forms by the dozen or so jet streams coursing through its enormous atmosphere. Much delicate microstructure could be discerned with careful study. "You've got to take a look at this," I cried out, as I heard my wife's approaching footsteps grow louder in my eardrum. Swinging the image to the eastern edge of the field, I arose from my observing chair and gestured to her to get comfortable behind the eyepiece. Now, my wife is no stranger to the delights presented by a telescopic image, but I could tell immediately she was gobsmacked by the Jovian apparition that was, by now, floating through the center of the orthoscopic field. "Gorgeous! Absolutely gorgeous!" she exclaimed, followed by a deafening silence as she cozied up to drink in the view.

Sharing a hot flask, we continued to observe Jupiter for another half hour or so. Then, the sacred covenant between observer and sky was broken. The wind got up and a few clouds rolled in, slowly at first, but then, faster and in ever growing numbers, coalescing swiftly into one monolithic form, engulfing the starry heavens. With effortless disregard for our petty concerns, the clouds extinguished those glorious details revealed just minutes before.

All good things must come to an end, I suppose. Yet, for a brief spell on that dark evening, my wife and I experienced what that great sage, William Butler Yeats, had prophetically written about all those years ago:

Once out of nature I shall never take
My bodily form from any natural thing
But such a form as Grecian goldsmiths make
Of hammered gold and gold enamelling
To keep a drowsy Emperor awake
Or set upon a golden bough to sing
To the Lords and ladies of Byzantium
Of what is past, or passing, or to come.

One of the author's magic flutes (Image by the author)

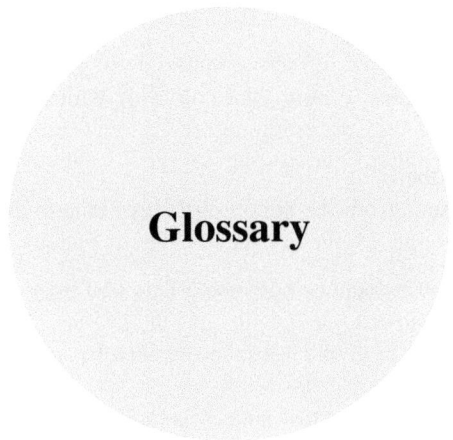

Glossary

Airy disk   The disk into which the image of a star is spread by diffraction in a telescope. The size of the Airy disk limits the resolution of a telescope

Alt-azimuth   A type of mount, like a simple photographic mount that allows you to make simple movements from left to right (azimuth) and up and down (altitude)

Antireflection coating   The application of a very thin layer of a substance (e. g., magnesium fluoride) to the surface of the lens, which has the effect of increasing light transmission and reducing internal reflections in the glass

Astigmatism   An aberration that occurs when there is a difference in the magnification of the optical system in the tangential plane and that in the Sagittal plane

ATMer   Amateur telescope maker

Barlow lens   A concave achromatic lens with negative focal length, used to increase the magnification of a telescope

Brashear process   A chemical method of depositing a thin layer of silver on a glass substrate; named after its developer, John A. Brashear

Coefficient of thermal expansion (CTE)   This measures the degree of expansion or contraction for each degree of temperature change. The lower the CTE, the less the material changes its shape while experiencing a temperature change

Collimation   The process of ensuring that all the optical elements in a Newtonian are perfectly in line with each other for maximum performance

Coma   An aberration that causes a point object to be turned into a pear- or comet-shaped geometry at the focal plane, and which most commonly manifests itself off-axis

Dauguerreotype   The earliest type of permanent photographic image usually produced on a silvered copper plate

N. English, *Classic Telescopes: A Guide to Collecting, Restoring, and Using Telescopes of Yesteryear*, Patrick Moore's Practical Astronomy Series, DOI 10.1007/978-1-4614-4424-4, © Springer Science+Business Media New York 2013

Depth of focus   A measure of how easy it is to attain and maintain a sharp focus. The larger the focal ratio of your 'scope, the greater its focus depth

Diffraction   A wave phenomenon that occurs when waves bend or distort as they pass around an obstacle

Dispersion   The tendency of refractive materials (e.g., a lens or prism) to bend light to differing degrees, causing the colors of white light to separate into a rainbow of colors

Extrafocal   Outside focus

Eye relief   The distance from the vertex of the eye lens to the location of the exit pupil

Focal length   The linear distance between a lens and the point at which it brings parallel light rays to a focus

Focal ratio   The focal length of a telescope divided by its aperture. Often denoted by f

Fresnel rings   The set of diffraction rings seen around stars just outside and inside focus

Intrafocal   Inside focus

Magnification   The factor by which a telescope makes an object larger

Multi-coated   The lenses are antireflection coated with more than one layer of coatings

Object glass   An older name for the glass objective of a refracting telescope

Parabolic mirror   The main mirror in a reflecting telescope that usually has a parabolic shape

Rayleigh criterion   In 1878 Lord Rayleigh concluded that the image of a point source such as a star would not be significantly impaired if the optical path difference from all the aberrations did not exceed a quarter of the wavelength of the light used to form the image. The quarter wave standard is often referred to as diffraction limit

Relative aperture   An old name for f ratio

Ronchi test   A method of determining the surface shape/figure of a mirror used in reflecting and other types of telescopes using a screen on which there are a large number of finely etched lines

Sacek effect   The ability of a classical achromat to hold greater displaced energy within the Airy disk owing to its large depth of focus. Named in honor of its discoverer, Vladimir Sacek

Secondary mirror   This is the flat mirror that directs light from the primary mirror of the telescope into the eyepiece

Secondary spectrum   The unfocused light, usually a purple fringing, produced by standard achromatic refractors

Speculum metal   An alloy of copper and tin and/or a small amount of other elements used for the construction of metal mirrors in Newtonian reflectors

Spherical aberration The inability to focus rays of light emanating from the center and edges of a lens at a single point in the image plane

Spherical mirror A mirror with a spherical figure usually only found in smaller reflecting telescopes

Spherochromatism The variation of spherical aberration with the wavelength of light

Strehl ratio A measure of optical quality that measures how much an optic deviates from perfection. A Strehl ratio of 1.0 is the best one can attain

Turned edge An aberration that occurs when the edge does not end abruptly but curls over gradually, starting from about 80% of the way out from the center of the mirror

Zerodur A very low expansion ceramic used as a substrate for making telescope mirrors

Zonal errors Localized defects that arise during the figuring and polishing of optical mirrors

References
and
Bibliography

Books

Argyle B (2007) The Observatory 127:392–400

Bell L (1981) The Telescope. Dover

Bennet A (1987) The divided circle, A history of instruments for astronomy navigation and surveying. Christies, England

Brashear JA (1988) *A man who loved the stars*: The autobiography of John Alfred Brashear. University of Pittsburgh Press

Chapman A (1999) The Victorian Amateur Astronomer: Independent Astronomical research in Britain 1820–1920, John Wiley & Sons.

Couteau P (1981) Observing visual double stars. M.I.T Press

English N (2011) A glass act. Astronomy Now 25(2):38–41

English N (2010) Choosing and using a refracting telescope. Springer

Hall AR (1992) Isaac Newton, Adventurer in thought. Cambridge University Press

King HC (1955) The history of the Telescope. Dover

Manly P (1994) The 20cm Schmidt Cassegrain telescope. Cambridge University Press

Manly P (1995) Unusual Telescopes. Cambridge University Press

McConnell A (1992) Instrument makers to the world: history of cook, troughton and simms from 1750. Ebor Press

Sheehan W (1995) The Immortal fire within- the life and work of Edward Emerson Barnard. Cambridge University Press

Sheehan W, O' Meara J (2001) Mars; the lure of the red planet. Prometheus Books

Sidgwick JB (1971) Amateur Astronomer's Handbook. Faber and Faber

Warner DJ (1968) Alvan Clark & Sons, Artists in optics. Smithsonian Institution Press

N. English, *Classic Telescopes: A Guide to Collecting, Restoring, and Using Telescopes of Yesteryear*, Patrick Moore's Practical Astronomy Series, DOI 10.1007/978-1-4614-4424-4, © Springer Science+Business Media New York 2013

Online Resources

The Antique Telescope Society. http://webari.com/oldscope/

Trudy Bell's essay on historic telescopes and their value to the society. http://www.capjournal.org/issues/09/09_30.pdf

Peter Abraham's historic telescope website. http://www.europa.com/~telscope/binotele.htm

Cloudy nights classic telescope forum, a place to discuss classic and or antique telescopes. http://www.cloudynights.com/ubbthreads/postlist.php/Cat/0/Board/classics

Details of the work of H. Dennis Taylor, master optician to T. Cooke & Sons, York. http://www.europa.com/~telscope/hdtaylor.txt

A list of some of the more obscure American telescope makers. http://home.europa.com/~telscope/tsusobs.txt

Useful information on John A. Brashear, an American telescope maker, can be found here. http://adsabs.harvard.edu/full/1920PA.....28..373S

More on Tom cave. http://www.mars.dti.ne.jp/~cmo/198/cave.html

The Unitron story. http://www.company7.com/library/unitron/unitron_114.html

The Criterion RV-6. http://www.company7.com/library/criterion_rv6.html

Robert Provin's excellent online resource for classic scope manuals. http://geogdata.csun.edu/~voltaire/classics/

A peek back in time to the workmanship of William Mogey & Sons. http://geogdata.csun.edu/~voltaire/classics/mogey/1932mogey.pdf

Binary star measurements with a seventeenth century, long-focal, non-achromatic refractor. http://www.jdso.org/volume6/number4/binder47.pdf

Some technical background on the achromatic refractor to be found on Vladimir Sacek's excellent online resource. http://www.telescope-optics.net/polychromatic_psf.htm

More on the Fry telescope, ULO, and its restoration. http://www.ulo.ucl.ac.uk/telescopes/fry/

Martian observations with the fry refractor, ULO. http://www.ulo.ucl.ac.uk/images/mars2003/

A 5-inch Cooke meets a 5-inch D&G. http://www.cloudynights.com/item.php?item_id=742

English N (2010) An inconvenient truth concerning small classical refractors. Stranger than fiction. http://www.cloudynights.com/item.php?item_id=2529

Mobberley M, Goward K Will hay and his telescopes. http://martinmobberley.co.uk/WillHay.html

The astronomical adventures of C. V. Raman. http://www.ias.ac.in/currsci/25oct2010/1127.pdf

More on the Hampstead Scientific Society. http://www.hampsteadscience.ac.uk/

The Story of Zeiss. http://www.company7.com/zeiss/history.html

A Zeiss telescope catalog from the 1920s. http://geogdata.csun.edu/~voltaire/classics/zeiss/zeiss1.pdf

The life and times of Dmitri Maksutov. http://www.telescopengineering.com/history/DmitriMaksutov.html

Traditional telescope making at its best – Dick Parker's excellent website. http://mirrorworkshop.mtbparker.com/refractorStory.html

The life and times of George Calver. http://www.oasi.org.uk/Hist_Ast/Calver.htm

A short discourse on the development of the apochromatic refractor. http://rohr.aiax.de/chapter%204a.htm

Index